T0140384

Compendium of Plant Genomes

Series editor

Chittaranjan Kole, Raja Ramanna Fellow, Department of Atomic Energy, Government of India, Kalyani, Nadia, West Bengal, India

Whole-genome sequencing is at the cutting edge of life sciences in the new millennium. Since the first genome sequencing of the model plant Arabidopsis thaliana in 2000, whole genomes of about 70 plant species have been sequenced and genome sequences of several other plants are in the pipeline. Research publications on these genome initiatives are scattered on dedicated web sites and in journals with all too brief descriptions. The individual volumes elucidate the background history of the national and international genome initiatives; public and private partners involved; strategies and genomic resources and tools utilized; enumeration on the sequences and their assembly; repetitive sequences; gene annotation and genome duplication. In addition, synteny with other sequences, comparison of gene families and most importantly potential of the genome sequence information for gene pool characterization and genetic improvement of crop plants are described.

Interested in editing a volume on a crop or model plant? Please contact Dr. Kole, Series Editor, at ckole2012@gmail.com

More information about this series at http://www.springer.com/series/11805

Ajit Kumar Shasany · Chittaranjan Kole
Editors

The *Ocimum* Genome

Editors
Ajit Kumar Shasany
Department of Plant Biotechnology
Council of Scientific and Industrial Research-Central Institute of Medicinal and Aromatic Plant Research (CSIR-CIMAP)
Lucknow, India

Chittaranjan Kole
Raja Ramanna Fellow, Department of Atomic Energy
Government of India
Kalyani, Nadia, West Bengal
India

ISSN 2199-4781 ISSN 2199-479X (electronic)
Compendium of Plant Genomes
ISBN 978-3-030-07355-8 ISBN 978-3-319-97430-9 (eBook)
https://doi.org/10.1007/978-3-319-97430-9

This Springer imprint is published by the registered company Springer Nature Switzerland AG
The registered company address is: Gewerbestrasse 11, 6330 Cham, Switzerland

This book series is dedicated to
my wife Phullara, and our children
Sourav, and Devleena

Chittaranjan Kole

Foreword

Members of the genus *Ocimum* are endowed with a wide array of aroma and healing properties. Commonly known as "Tulasi," the "holy basil" (*Ocimum tenuiflorum*) is revered and worshiped throughout India. It is one of the most important plants of the traditional systems of Indian medicine, which has been used extensively in various Ayurvedic and Yunani formulations for the treatment of various health problems. Due to its medicinal and religious importance, it has also been described as "Mother Medicine of Nature," "The Incomparable One," and "The Queen of Herbs" by different authors. Tulasi acts as a stimulant for physical, intellectual, and spiritual activities, which addresses many modern-day health issues, and is an integral part of the Ayurveda's holistic lifestyle approach to health. It is believed that daily intake of Tulasi supports immunity, averts infections and diseases, promotes general health, longevity, and well-being, and also helps in dealing with the stresses of everyday life. Despite being such an important plant, the molecular studies defining the pathways and plant metabolites responsible for its diverse aroma and therapeutic activities started very late. The next-generation sequencing of the whole genome of *Ocimum* has introduced this wonder herb to the modern science and opened up the doors for unraveling its medicinal secrets, which will pave the way to harness the therapeutic potential of this holy herb.

The essential oils of the genus *Ocimum* find various applications in fragrance and cosmetic industry as well as local medicinal systems of different countries of the world. With its ever-increasing demand from aromatic industry, the concerns over improving productivity and quality of its raw materials are also increasing. Probing the transcriptome and the whole-genome sequence of *Ocimum* can help in identifying molecular markers for plant breeding to facilitate the development of improved varieties with the desired oil composition. The genus *Ocimum* like its usefulness is very wide and diverse. Chemo-profiling and pharmacognostic research followed by in vivo validation of medicinal properties have been the major focus of interest in *Ocimum* research for long, but with the advent of next-generation sequencing "*Ocimum* genomics" is fast emerging and progressing. The book *The Ocimum Genome* has been compiled with an aim to portray some of the most recent developments in *Ocimum* breeding and genomics. I highly appreciate the hard work of the editors in organizing the diverse information and perception of the distinguished groups of scientists in a concise manner within ten chapters.

I am delighted to compliment the editors, Prof. Chittaranjan Kole, an internationally renowned agricultural scientist with original contributions on crop genomics, and Dr. Ajit Kumar Shasany, Chief Scientist at CSIR-CIMAP, an eminent scientist in the area of medicinal and aromatic plant research, and the anchor person behind the "*Ocimum* genome story" for bringing out a much desired book *The Ocimum Genome*. I am confident that this book would be a very useful resource of knowledge to the scholars and scientists, not only working on *Ocimum* but also on other medicinal or aromatic plants. Besides, this book is also expected to serve science managers and policy makers.

I sincerely hope that the present attempts to understand the hidden potential of *Ocimum* with modern approaches and outlook will open many new opportunities.

Lucknow, India

Anil Kumar Tripathi
Director, Council of Scientific and Industrial
Research-Central Institute of Medicinal
and Aromatic Plant Research (CSIR-CIMAP)

Preface to the Series

Genome sequencing has emerged as the leading discipline in the plant sciences coinciding with the start of the new century. For much of the twentieth century, plant geneticists were only successful in delineating putative chromosomal location, function, and changes in genes indirectly through the use of a number of 'markers' physically linked to them. These included visible or morphological, cytological, protein, and molecular or DNA markers. Among them, the first DNA marker, the RFLPs, introduced a revolutionary change in plant genetics and breeding in the mid-1980s, mainly because of their infinite number and thus potential to cover maximum chromosomal regions, phenotypic neutrality, absence of epistasis, and codominant nature. An array of other hybridization-based markers, PCR-based markers, and markers based on both facilitated construction of genetic linkage maps, mapping of genes controlling simply inherited traits, and even gene clusters (QTLs) controlling polygenic traits in a large number of model and crop plants. During this period, a number of new mapping populations beyond F2 were utilized and a number of computer programs were developed for map construction, mapping of genes, and for mapping of polygenic clusters or QTLs. Molecular markers were also used in studies of evolution and phylogenetic relationship, genetic diversity, DNA-fingerprinting, and map-based cloning. Markers tightly linked to the genes were used in crop improvement employing the so-called marker-assisted selection. These strategies of molecular genetic mapping and molecular breeding made a spectacular impact during the last one and a half decades of the twentieth century. But still they remained 'indirect' approaches for elucidation and utilization of plant genomes since much of the chromosomes remained unknown and the complete chemical depiction of them was yet to be unraveled.

Physical mapping of genomes was the obvious consequence that facilitated development of the 'genomic resources' including BAC and YAC libraries to develop physical maps in some plant genomes. Subsequently, integrated genetic–physical maps were also developed in many plants. This led to the concept of structural genomics. Later on, emphasis was laid on EST and transcriptome analysis to decipher the function of the active gene sequences leading to another concept defined as functional genomics. The advent of techniques of bacteriophage gene and DNA sequencing in the 1970s was extended to facilitate sequencing of these genomic resources in the last decade of the twentieth century.

As expected, sequencing of chromosomal regions would have led to too much data to store, characterize, and utilize with the-then available computer software could handle. But development of information technology made the life of biologists easier by leading to a swift and sweet marriage of biology and informatics, and a new subject was born—bioinformatics.

Thus, evolution of the concepts, strategies, and tools of sequencing and bioinformatics reinforced the subject of genomics—structural and functional. Today, genome sequencing has traveled much beyond biology and involves biophysics, biochemistry, and bioinformatics!

Thanks to the efforts of both public and private agencies, genome sequencing strategies are evolving very fast, leading to cheaper, quicker, and automated techniques right from clone-by-clone and whole-genome shotgun approaches to a succession of second generation sequencing methods. Development of software of different generations facilitated this genome sequencing. At the same time, newer concepts and strategies were emerging to handle sequencing of the complex genomes, particularly the polyploids.

It became a reality to chemically—and so directly—define plant genomes, popularly called whole-genome sequencing or simply genome sequencing.

The history of plant genome sequencing will always cite the sequencing of the genome of the model plant Arabidopsis thaliana in 2000 that was followed by sequencing the genome of the crop and model plant rice in 2002. Since then, the number of sequenced genomes of higher plants has been increasing exponentially, mainly due to the development of cheaper and quicker genomic techniques and, most importantly, development of collaborative platforms such as national and international consortia involving partners from public and/or private agencies.

As I write this preface for the first volume of the new series 'Compendium of Plant Genomes,' a net search tells me that complete or nearly complete whole-genome sequencing of 45 crop plants, eight crop and model plants, eight model plants, 15 crop progenitors and relatives, and three basal plants is accomplished, the majority of which are in the public domain. This means that we nowadays know many of our model and crop plants chemically, i.e., directly, and we may depict them and utilize them precisely better than ever. Genome sequencing has covered all groups of crop plants. Hence, information on the precise depiction of plant genomes and the scope of their utilization is growing rapidly every day. However, the information is scattered in research articles and review papers in journals and dedicated Web pages of the consortia and databases. There is no compilation of plant genomes and the opportunity of using the information in sequence-assisted breeding or further genomic studies. This is the underlying rationale for starting this book series, with each volume dedicated to a particular plant.

Plant genome science has emerged as an important subject in academia, and the present compendium of plant genomes will be highly useful both to students and teaching faculties. Most importantly, research scientists involved in genomics research will have access to systematic deliberations on the plant genomes of their interest. Elucidation of plant genomes is of interest not only for the geneticists and breeders, but also for practitioners of an array of plant science disciplines, such as taxonomy, evolution, cytology, physiology, pathology, entomology, nematology, crop production, biochemistry, and obviously bioinformatics. It must be mentioned that information regarding each plant genome is ever-growing. The contents of the volumes of this compendium are therefore focusing on the basic aspects of the genomes and their utility. They include information on the academic and/ or economic importance of the plants, description of their genomes from a molecular genetic and cytogenetic point of view, and the genomic resources developed. Detailed deliberations focus on the background history of the national and international genome initiatives, public and private partners involved, strategies and genomic resources and tools utilized, enumeration on the sequences and their assembly, repetitive sequences, gene annotation, and genome duplication. In addition, synteny with other sequences, comparison of gene families, and, most importantly, potential of the genome sequence information for gene pool characterization through genotyping by sequencing (GBS) and genetic improvement of crop plants have been described. As expected, there is a lot of variation of these topics in the volumes based on the information available on the crop, model, or reference plants.

I must confess that as the series editor, it has been a daunting task for me to work on such a huge and broad knowledge base that spans so many diverse plant species. However, pioneering scientists with lifetime experience and expertise on the particular crops did excellent jobs editing the respective volumes. I myself have been a small science worker on plant genomes since the mid-1980s and that provided me the opportunity to personally know several stalwarts of plant genomics from all over the globe. Most, if not all, of the volume editors are my longtime friends and colleagues. It has been highly comfortable and enriching for me to work with them on this book series. To be honest, while working on this series I have been and will remain a student first, a science worker second, and a series editor last. And I must express my gratitude to the volume editors and the chapter authors for providing me the opportunity to work with them on this compendium.

I also wish to mention here my thanks and gratitude to the Springer staff, Dr. Christina Eckey and Dr. Jutta Lindenborn in particular, for all their constant and cordial support right from the inception of the idea.

I always had to set aside additional hours to edit books besides my professional and personal commitments—hours I could and should have given to my wife, Phullara, and our kids, Sourav, and Devleena. I must mention that they not only allowed me the freedom to take away those hours from them but also offered their support in the editing job itself. I am really not sure whether my dedication of this compendium to them will suffice to do justice to their sacrifices for the interest of science and the science community.

Kalyani, India Chittaranjan Kole

Preface

The genus *Ocimum* L. is a member of family Lamiaceae, collectively called basil. The distinctive characteristics in this family are a square stem with leaves arranged in opposite decussate manner having many gland dots. The flowers have two distinct lips and are strongly zygomorphic. Many researchers have observed complexity in their chromosome numbers and ploidy levels. As per the chromosome count database (CCDB), the chromosome count of the genus ranges from a minimum of $(n) = 6$ to the maximum of $(n) = 64$. The name basil is derived from the Greek word *basilikos*, which means royal or the king. There are around 35–150 species included in the genus *Ocimum*, which are annual and perennial herbs/shrubs indigenous to Africa, Asia, Central, and South America, but extensively disseminated worldwide. The finest known strongly aromatic species are the herbs *Ocimum gratissimum* (African basil) and *Ocimum basilicum* (Thai basil) which have a pleasant odor due to the essential oils consisting of monoterpenes, sesquiterpenes, and phenylpropanoids. But the genus is also popular for its medicinal properties and spiritual sanctity which are mainly acquired by the herb *Ocimum sanctum*, also called *Ocimum tenuiflorum* (holy basil or *tulsi* in Hindi). Extracted essential oils have also been shown to contain biologically active constituents that are insecticidal, nematicidal, and fungistatic. Such properties are often attributed to dominating essential oil constitutes such as eugenol, methyl chavicol, camphor, linalool, and methyl cinnamate. Three main forms of *Ocimum tenuiflorum* are generally popular, one is *Rama tulsi* with green stems and leaves, other is *Krishna tulsi* with purple colored stems and leaves, and the third one is *Vana tulsi* which is the basic wild form. Though *Ocimum tenuiflorum* is one of the most revered plants in India having the miraculous medicinal properties, there is very less work carried out to know about the biochemical pathways of the miracle compounds which are undoubtedly the secondary metabolites present in the plant. However, in the past 5 years there had been a great pace in research on the transcriptomics and the functional genomics of *Ocimum* sp. in addition to the whole-genome sequencing of the holy basil (*Ocimum tenuiflorum*). This would help enabling the discovery of genes and the molecular markers for plant breeding to develop new improved varieties.

This book covers major aspects in the *Ocimum* spp. research in ten chapters in an attempt to integrate contemporary research efforts and highlighting some of the most exhilarating advances in *Ocimum* breeding and genomics.

Chapter 1 focuses upon the agro-technological practices and field requirements for growing the basil plant and its economics covering the cost input to output ratio. This chapter not only describes the techniques of successful cultivation of basil, but also gives an overview of earning by farmers by growing the basil crop.

Chapter 2 highlights the medicinal and therapeutic potential of the various *Ocimum* species with the main emphasis on the antioxidant and antiaging potential of phytomolecules and extracts derived from this genus. Since basil is the rich repository of many bioactive molecules such as terpenes, phenylpropenes, phenolic acids, and flavonoids, therefore the longevity-promoting potential of this herb is highlighted in the present chapter. Furthermore, special focus is given to the use of *C. elegans* model system for screening various phytochemicals isolated from the *Ocimum* spp.

Chapter 3 discusses the profound effect of *Ocimum* on treatment and prevention of cardiovascular diseases by means of lowering blood lipid content, suppressing ischemia and stroke, reducing hypertension, and also its higher antioxidant properties. These cardioprotective properties prove that *Ocimum* may be treated as a good remedy against prevention and treatment of cardiovascular diseases.

Chapter 4 demonstrates the evolutionary analysis of few protein superfamilies in *Ocimum tenuiflorum.* This chapter describes a computational pipeline for identification, validation, and analysis of the key components involved in the synthesis of terpenoids and a less studied class of proteases called rhomboids. This kind of study will have wider implication not only as a tool to understand sequence and structure–function relationships of some of the well-studied metabolites and enzymes, to aid protein engineering for biotechnological utilization of these highly commercially valuable molecules.

Chapter 5 presents the detailed taxonomic description and phylogenetic relationships among the Indian *Ocimum* species. Chapter 6 highlights the genetics, cytogenetics, and diversity in the genus *Ocimum* and the ambiguities in classification of the genus. It elaborates that how the revision in nuclear DNA content has divided the section *Ocimum* into two clades. The first consisting of *O. basilicum* and *O. minimum*, whereas the second consisting of *O. americanum, O. africanum*, and two *O. basilicum* var. *purpurascens* accessions. Out of all, *O. tenuiflorum* was found to be the most divergent species.

Chapter 7 emphasizes on the traditional plant breeding carried out in developing new improved *Ocimum* varieties like varieties CIM Ayu, CIM Angana, CIM Saumya, CIM Kanchan, Vikarsudha, CIM Jyoti, CIM Sharada, CIM Surabhi, and CIM Snigdha developed at the CSIR-Central Institute of Medicinal and Aromatic Plants, Lucknow, India.

Chapter 8 describes all the available genomic resources of the *Ocimum* species worldwide and illustrates the medicinal potentialities, uses, and essential oil components of some of the widely used *Ocimum* species.

Chapter 9 discusses the triterpene functional genomics in *Ocimum via* utilization of high-throughput sequencing of genomes and transcriptomes providing a prospect leading to the understanding of the molecular and biochemical basis for the biosynthesis of diverse triterpenes and other phytochemicals in *Ocimum* species.

Chapter 10 gives an overview of the genome sequencing of the holy basil and the future prospects of utilizing the genome sequencing data.

We consciously tried to present these ten chapters as more or less stand-alone deliberations, and hence, one would find obviously some apparent redundancy while introducing the plant.

We hope that this book would be a great help to the senior and the young *Ocimum* researchers and enthusiastics in providing an insight about the present status of genomic, pharmacognostic, agronomic, and the evolutionary studies. This compilation not only projects massive prospective for plant improvement but also unveils the new avenues which were obscured from long. We are really grateful to each of the authors for sparing their precious time and effort in order to contribute their best. We would also like to thank the entire editorial team of Springer for their assistance.

Lucknow, India
Kalyani, India

Ajit Kumar Shasany
Chittaranjan Kole

Contents

Ocimum as a Promising Commercial Crop

1

R. K. Srivastava, Sanjay Kumar and R. S. Sharma

Abstract

Basil, *Ocimum* Spp, (Family Labiatae), is a herbaceous, erect, annual important aromatic plant, which attains the height of about 80–100 cm. The leaves of the plant are dark green or yellowish green in colour. Flowering tops and leaves of plant yields essential oils, which are used in perfumery and pharmaceutical industries. Stems of the plant are often branched and bear leaves. Many species of basil are available in nature including *Ocimum basilicum, Ocimum canum, Ocimum gratissimum* and *Ocimum sanctum* (Sharma et al. in J Med Arom Plant Sci 18:512–522, 1996). Basil can be grown in wide ranges of soil like light loam and medium loam having good water holding capacity with a pH range of 5.0–8.30. The best crop rotation of basil is basil–chamomile–mint or basil–mustard–mint or basil–potato–mint in the subtropical region. It prefers mild climate with moderate temperature of about 27 °C for successful growth. Basil is propagated through seeds. The nursery is raised in the month of May, and the seedlings are transplanted in the main field in the month of June/July. Nursery-raised seedlings of 30 days' age are planted with the spacing of 30–35 cm plant to plant and 45–50 cm row to row depending upon soil fertility. After planting of seedlings, irrigation is necessary. During the whole period of life, 2–3 weedings are required to minimize weed competition. In average fertile soil, 50 kg nitrogen, 40 kg phosphorus and 40 kg potash per ha are sufficient. Nitrogen is applied in three equal doses during the growth period of the plant. It takes about 85–90 days for maturity, when lower leaves start turning yellow and full blooming condition appears. Harvesting is done by sharp sickle. After harvesting and distillation, about 110 kg of oil is received from per hectare area. The present market rate of basil oil is Rs. 650 per kg (Essential oil Market Report 2014), and cost of cultivation is about Rs. 23,546 to per ha. A farmer can earn Rs. 47,954 per ha within a period of 100 days.

1.1 Cultivation Technology

Basil is known for its leaves, roots, stem and essential oils for fragrances, flavours, medicine and sanctity in Hindu mythology. A number of species are available in the world, but there are two common species that are more popular among the farmers, i.e. *Ocimum sanctum* (holy basil) and *Ocimum basilicum* (sweet basil).

R. K. Srivastava (✉) · S. Kumar · R. S. Sharma
CSIR-Central Institute of Medicinal and Aromatic Plants (CIMAP), Near Kukrail Picnic Spot, Lucknow 226015, India
e-mail: rksrivastava@cimap.res.in

A. K. Shasany and C. Kole (eds.), *The Ocimum Genome*, Compendium of Plant Genomes, https://doi.org/10.1007/978-3-319-97430-9_1

The leaves of this plant or whole plant are used for extraction of its essential oils. The oil contains mainly phenols, aldehydes, tannins, saponin and fats (Smitha et al. 2014). About 150 species of *Ocimum* have been found in the world, out of which some are wildly grown in India. A few are commercially cultivated for their important essential oil and crude drug purpose, but the sweet basil or French basil is cultivated for its essential oil.

Apart from the sweet basil, some of the holy basil is also cultivated for their chemical constituents but the sweet basil is widely cultivated in India and in other parts of the world. The sweet basil is native of India, but it is also cultivated in France, Italy, Bulgaria, Egypt, Hungary, South Africa and USA. The oil of this species contains methyl chavicol, linalool and some other minor constituents (Srivastava et al. 2009; Fig. 1.1).

O. basilicum is erect, herbaceous, strongly aromatic, annual plant having a height of 75–95 cm. Stem often branched from the base, purplish and subglabrous in lower part, four angular, intensely hairy higher up. Leaves are 3–6 cm long and 1.5–3.5 cm in diameter, ovate, lanceolate, with acuminate or rounded base, acute or subobtuse, subentire to serrate, thinly hairy and gland pinnate. Flowers are born in long terminal racemose inflorescences. Pedicles densely hairy, calyx 0.3–0.4 cm long at first, afterwards 0.5–0.6 cm long hairy outside, glabrous inside, at base with dense whorl of long hairs above it, glandulose upper tip ovate to round. Corolla white or pale purple tube, stamens exert, posterior filaments short, white hairy and transverse (Srivastava et al. 2009).

O. basilicum and *O. sanctum* are cultivated on large scale in Indian states. Mostly, they are cultivated for important essential oil in the states of Uttar Pradesh, Punjab, Haryana, Madhya Pradesh and some other parts of India.

1.2 Common Varieties

Ocimum basilicum

CIM-Saumya: This is a short duration crop of 90 days and has the potential to produce about 85–110 kg/ha essential oil and rich in methyl chavicol (62%) and linalool (25%) (Bahl et al. 2018).

CIM-Snigdha: The variety matures in 80–90 days and yields 75–110 kg per ha essential oil and also rich in methyl cinnamate content (78–80%) (Bahl et al. 2018).

CIM-Surabhi: This is high oil-yielding variety with unique chemical composition having 70–75% (−) linalool with 99% purity. This

Fig. 1.1 A field view of sweet basil

Classification:

Kingdom: Plantae

Class: Angiosperm

Oder: Lamiales

Family: Lamiaceae

Genus: *Ocimum*

Species: *basilicum* L.

variety yields 100–120 kg/ha essential oils within 100 days (Bahl et al. 2018).

Ocimum sanctum

CIM-Ayu: This variety is having the potential to produce 15–16 quintals dry leaf or yield 80–100 kg/ha oil rich in eugenol (80–83%) (Bahl et al. 2018).

CIM-Angna: This variety is having the capacity to produce dry leaf (12–14 q/ha) or 75–90 kg/ha essential oil containing eugenol (37–40%) and germacrene-D (15–16%) (Bahl et al. 2018).

1.3 Soil and Climate

The crops are grown well on moderate fertile and well-drained sandy loam to loam soils with good water holding capacity with a pH range of 5–8. It requires moderate climate (Temp 25–35 °C). North Indian plains are suitable for its cultivation. In hills, it can be cultivated only as a summer crop (Srivastava et al. 2009; Fig. 1.2).

1.4 Land Preparation

The field should be disc ploughed once and harrowed twice with the mixing of 10–15 tons of FYM per hectare. It should be ready for planting in June to first week of July.

1.5 Propagation

Basil crop can be propagated through nursery and direct sowing/broadcasting. However, direct sowing is not advisable for cultivation due to lower yield and higher rate of seed.

1.5.1 Raising of Nursery

The cultivation of Indian basil through raising of nursery is recommended. The nursery bed should be cleaned of stubbles, weeds and soil nematodes; 1.0–1.5 tons FYM, 15 kg nitrogen, 20 kg phosphorous and 10 kg potash should be applied in nursery bed of 500 m^2. About 500 m^2 nursery area would be required for one ha plantation. The size of beds should be 2 × 4 m with proper

Fig. 1.2 A view of sweet basil after two months

irrigation channel. As the seeds are small, they should be mixed with 5–8 times sand or fine dried soil. Nearly 800 g–1.5 kg good germinated seeds are required for raising nursery for one hectare area. Appropriate time of raising nursery is last week of May or first week of June in North Indian conditions. Sowing should be done in line 5–10 cm apart, and care should be taken to avoid deep sowing. It is advisable to cover the seed beds lightly with straw so as to conserve the moisture which may be removed after germination of the seeds. In dry season, it may be necessary to irrigate the nursery beds twice a day by light irrigation. Seedlings are ready for transplantation within 25–30 days (Srivastava et al. 2009).

1.5.2 Transplanting

Transplanting should be done in the morning or evening hours; just after the planting, irrigation is required. Cloudy weather and light rain are considered ideal for transplanting. It is recommended to transplant the seedlings in rows and plant spacing of 50 × 40 cm or 50 × 30 cm, respectively, depending upon soil fertility (Bahl et al. 2018).

1.5.3 Direct Sowing

Field should be cleaned of stubbles and weeds; about 10–15 tons FYM is required during preparation of the land. In large-scale cultivation, farmers cultivate the Indian basil through broadcasting method. In this type, about 2.5–3.0 kg seeds are required for 1 ha of land. One part of the seed and ten parts of the sand should be mixed for proper broadcasting. By broadcasting, spacing of the plants cannot be maintained. The thinning is advisable during first weeding for maintaining the population density of the plants (Bahl et al. 2018; Fig. 1.3).

1.5.4 Irrigation

Irrigation depends upon the moisture in the field. During monsoon, water requirement of the crop is fully met up to September; thereafter, 3–4

Fig. 1.3 A view of cultivation of basil in mango orchards

irrigations are required. However, in rainy season, no irrigation is required.

1.6 Interculture and Weeding

The first weeding should be done after 30–40 days of transplantation. The second weeding/hoeing is recommended after 1 month from first weeding, till plants become bushy to suppress the lower weeds.

In broadcasting type of sowing, weeding is required after 25–30 days from sowing. In first weeding, uprooting is required to unwanted plants to maintain proper population. The second weeding is recommended after 1 month of first weeding (Bahl et al. 2018).

1.7 Manures and Fertilizers

Essential oil, the product to be obtained, is synthesized in the leaves and inflorescence of basil. Thus, adequate fertilizers are essential for luxuriant growth with good leaves and inflorescence in the crops. In soil of average fertility, about 50–60 kg nitrogen, 40 kg phosphorus and 40 kg potash per hectare are advisable. About 20 kg nitrogen and total quantity of phosphorus and potash are applied as a basal dose at the time of planting. Rest 30 kg nitrogen is applied in two split doses during the growing season. Because it is a short duration crop, it is most responsive to nitrogen fertilizers (Srivastava et al. 2009).

1.8 Pest and Diseases

Basil crops are less affected by pests and diseases. The common diseases are Fusarium wilt caused by *Fusarium oxysporum*. During the preparation of nursery, Pythium damping off may also occur that damages the seedling; it is caused by *Pythium insidiosum*.

The basil crop is also affected by leaf rollers which stick to the under surface of the leaves, fold them backwards lengthwise and web them together.

Another pest that affects the basil crops is lace wing caused by *Cochlochi labullita*. The nymphs and adults eat leaves and younger stems, the leaves initially get curled, and later the whole plant gets dried up. It can be controlled by spraying of Azadirachtin 10,000 ppm @ 5 ml/l to control this insect (Smitha et al. 2014).

Basil is susceptible to powdery mildew (*Oidium* spp.), seedling blight (*Rhizoctonia solani*) and root-rot (*Rhizoctonia bataticola*). The powdery mildew disease is controlled by spraying of wettable sulphur (4 g/litre of water), and the latter two diseases can be managed by improved phytosanitary measures and by drenching the nursery beds with Bavistin 1% (Smitha et al. 2014).

1.9 Crop Rotation

Due to high productivity and low returns from cereal crops, agriculture should be diversified from traditional crops to high-value crops like medicinal and aromatic plants. Cultivating Indian basil as a bonus crop may be helpful in minimizing the expenses on cereal crops. The rotations are possible both with transplanted and direct planted basil crops. The following crop rotations are recommended:

Basil—Potato—Mint; Basil—Potato; Basil—Mint; Basil–*Brassica*—Mint; Basil–Pea—Mint; Basil–Wheat—Mint, Basil—Chamomile—Mint (Srivastava et al. 2009).

1.10 Harvesting

The basil crop herbage should be harvested when the field is dry and weather is sunny. The crop is harvested at a time of full bloom when lower leaves turn yellowish. The crop may come to full bloom in 2.5–3 months after transplanting at the subtropical region. In Northern India, only two harvests can be taken. In the first harvest, only flowering tops are harvested. But single harvest is more popular because the second harvest of basil takes more time and the proceeding cropping will be delayed (Dwivedi et al. 2000).

1.11 Distillation and Oil Yield

The distillation of the basil is done by hydro-steam distillation using field distillation technology. The yield of herbage from a basil crop depends upon a number of factors including the climate, soil properties, varieties used, timely plantation and irrigation, control of weeds, pests and diseases, fertilization, planting and harvesting schedule. Average herb yield of basil is about 20–25 tons, and oil yield is about 110–125 kg per hectare. For removing of any moisture, oil should be treated with anhydrous sodium sulphate and decanting (Srivastava et al. 2009).

1.12 Uses of Basil Oil

Essential oil of basil has a characteristic warm herbal anise-like note with slightly smoky background. Regarding flavour, it is sweet warm spicy anise-like with bitter aftertaste. It is used for imparting green herbal character in modern fragrances, particularly in chypre, floral aldehydic and also in modern fougere types. The general use level in such fragrances is up to 2–3%. It also used in other fragrances for cosmetics as well where herbal warm effect is required. In eau de cologne and green-type perfume, this oil gives very interesting effect. This oil also finds use in soap fragrances and in other low cost fragrance as well (Srivastava et al. 2009).

Basil oil is also a good source for certain isolates and semi-synthetic chemicals like linalool and anethol. Estragol (methyl chavicol) and linalool are present in this oil after conversion of estragol into anethol; the anethol and linalool are repented via fractionation. The anethol is used for imparting anise flavour, while linalool is used for flavour and fragrance both.

In aromatherapy, it is used as insect repellent and insect bites. It is good for bronchitis, cold, cough, earache, sinusitis, etc. The oil is antiseptic, antispasmodic, carminative, digestive, expectorant, prophylactive and insecticide in action. It is also used in fever, flue, anxiety, depression, fatigue and nervous tension (Srivastava et al. 2009; Table 1.1).

Table 1.1 Economics of cultivation of basil in India

Particulars per hectare	Cost of cultivation (Rs./ha/year)
A. Cost of cultivation and distillation	
Cost of seed (1 kg seed)	2500
Land preparation	2800
Fertilizer	2850
Labour cost (transplanting)	3000
Irrigation	900
Plant protection	700
Weeding (one time manually)	2400
Harvesting	3600
Transportation of herbs	1000
Distillation charges	3000
Interest on variable cost @ 7% interest	796
Total	23,546
B. Gross returns	
Oil yield (kg)	110
Price of oil (per kg)	650
Total cost of (oil per ha.)	71,500
C. Net returns	**47,954**

The above cost parameters are based on prevailing market rate of manpower, other materials, and produce (essential oil) may deviate in future.

References

Bahl JR, Singh AK, Lal RK, Gupta AK (2018) High-yielding improved varieties of medicinal and aromatic crops for enhanced income. In: Singh B, Peter KV (eds) New age herbals. Springer Nature Singapore Pte Ltd, pp 257–259

Dwivedi S, Mishra PN, Singh AP, Kothari SK, Naqvi AA, Kumar S (2000) Cultivation of sweet basil *Ocimum basilicum* in India. CIMAP Farm Bull 16:14

Market Report September (2014) Compiled and published by Indian Perfumer 58(3):15

Sharma JR, Ashok Sharma, Singh AK, Kumar S (1996) Economic potential and improved varieties of aromatic plants of India. J Med Arom Plant Sci 18:512–522

Smitha GR, Thania S, Manivel VP (2014) Technical bulletin of Basil, Directorate of Medicinal and Aromatic Plant Research. Anand, Gujarat, India

Srivastava RK, Shukla SV, Dagar SS (2009) Cultivation and uses of aromatic plants. IBDC Publishers, Lucknow, India

Ocimum Species: A Longevity Elixir

2

Aakanksha Pant and Rakesh Pandey

Abstract

Aging is a major risk factor associated with the period of morbidity and pain at a later stage of life. Although the average age for initiation of morbidity has delayed, chronic diseases like hypertension, cancer, diabetes, and neurodegenerative disorders are still prevalent in affluent aging societies leading to death. Therefore, studying dietary interventions and pathology of aging can prove as an essential strategy for achieving healthy aging. The recent researches demonstrated an association of aging with remarkable elevation in intracellular reactive oxygen species (ROS) and stress. The plant-based molecules have successfully modulated lifespan and stress level across various species. These phytomolecules are secondary plant metabolites which play a major role in plant defence network and are synthesized as side tracks of plant's primary metabolism. Despite the recently discovered potential of some phytomolecules in alleviating age-related stress, antiaging and stress modulatory potential of most of them is still unraveled. The natural dietary intervention modulating lifespan and health span in model organisms should be vastly studied for improving later life health. The free-living soil nematode, *Caenorhabditis elegans,* provides a unique and expedient platform for studying pharmacological interventions and dissecting the genetic mechanism underlying aging. The present chapter highlights the medicinal and therapeutic potential of the various *Ocimum* species which is commonly known as the "Queen of the herbs." The antioxidant and antiaging potential of phytomolecules and extracts derived from this genus is thoroughly reported. Basil is the rich repository of many bioactive molecules such as monoterpenes, sesquiterpenes, phenylpropanoids, anthocyanins, and phenolic acids; therefore, the longevity-promoting potential of this herb is highlighted in the present chapter. Furthermore, special attention is given to the employment of *C. elegans* model system for screening various phytochemicals isolated from *Ocimum* spp.

2.1 Aging

The milestone discoveries in medical sciences dramatically extended human lifespan over last few centuries of human history. The mounting percentage of the older population and incidence of age-related disorders are a huge

A. Pant · R. Pandey (✉)
Microbial Technology and Nematology Department, CSIR-Central Institute of Medicinal and Aromatic Plants, India Lucknow Rd, Neil Lines, Cantonment, Lucknow 226002, Uttar Pradesh, India
e-mail: r.pandey@cimap.res.in

A. K. Shasany and C. Kole (eds.), *The Ocimum Genome*, Compendium of Plant Genomes, https://doi.org/10.1007/978-3-319-97430-9_2

socioeconomic concern. The inevitable changes due to aging are universal to all organisms, and scientists have sought the key to youthfulness for centuries. Despite advancement in modern medical science, an elixir has so far proved elusive. Aging and age-related chronic disorders have become the central topic of political, medical, and scientific interest.

Aging is a debilitating process which reduces efficiency and function of an organism (Kenyon 2010). It is characterized by progressive deterioration of tissue integrity with age and decline in stress survival ability (Kirkwood 2005). The increment in age-related functional impairment can be attributed to developmental, genetic, or environmental factors which lead to the growing probability of death (Rattan 2006). The term "aging" depicts the passage of time, and the "senescence" explains all the processes of a lifetime from maturity to death (Garigan et al. 2002). The decline in reproduction, survival, and accumulation of metabolic by-products is the prime characteristic of senescence (da Costa et al. 2016). Aging affects efficiency and function of an organism, and the prime feature includes deterioration of skin, bones, blood vessels, lungs, other vital organs and nerves leading to loss of memory, motor coordination, hearing, and vision power in human beings (Longo et al. 2015). Aging is the single largest risk factor which influences various chronic ailments like cardiovascular, cancer, and neurodegenerative disorders (Blagosklonny 2009). Therefore, there is a need to understand and alleviate the age-related decline in health span and lifespan of an organism. Hence, the discovery of novel molecules, which modulate aging, could lead to a new strategy for working upon age-related diseases like diabetes, cancer, and neurodegenerative disorders (Fig. 2.1).

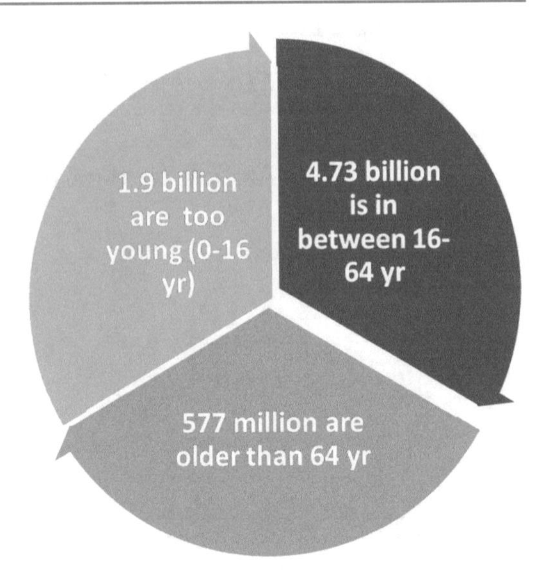

Fig. 2.1 Our aging world: about ≈ 577 million are older than 64 years out of approximately 7 billion world population (based on data. Source from CIA, World Bank, and gem consortium.org)

Although aging is a universal phenomenon, still it is poorly understood and the elderly population in the world is rising; hence, there is a great need to explore aging process and the genes/pathways defining it. Therefore, the interventions regulating aging which can reduce the severity of age-related pathologies promoting healthy lifespan in elderly have become more necessary than ever.

2.2 Theory of Aging

The question why we age is tantamount to what is life itself. There are countless theories explaining the aging phenomenon across organisms but until recent times the definition of aging—"cell senescence"—was uncertain. The development in the field of gerontology has generated multiple theories explaining the phenomenon of aging and its inevitable consequence, death. The complex process of aging is characterized by various changes that occur at different levels of the biological hierarchy. The search for the single key factor such as single gene or process has been replaced by the outlook considering aging as a complex multifactorial process. Multiple theories explaining aging are not mutually exclusive (Fig. 2.2) and may effectively elaborate the few or most of the characteristics of the aging process alone or as a cocktail with other theories (Rattan 2006). The theories explaining aging process have been

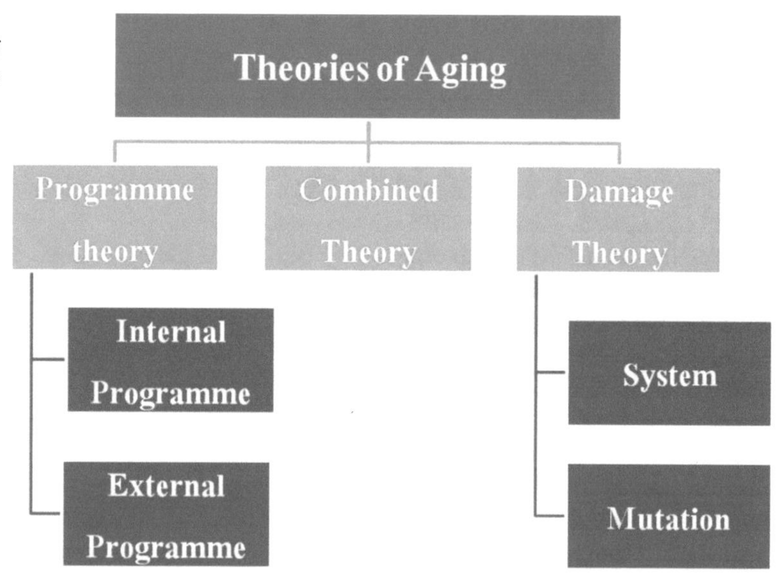

Fig. 2.2 Theories of aging [based on the classification of aging theories described by da Costa et al. (2016)]

grouped into two categories, programmed and error theories of aging (Rattan 2006).

The programmed theories suggest that aging depends on biological clocks regulating lifespan and the regulation of lifespan depends on genes monitoring nervous, endocrine, and immune system (Rattan 2006), whereas the error theories propose environmental insults inducing damage at the cellular level (mitochondrial dysfunction and accumulation of intracellular ROS) which leads to progression in the age in organisms (Kirkwood 2005). The most popular theory is primitive Harman's free radical theory which explains reactive oxygen species (ROS)-mediated oxidative stress and progression in age (Harman 1955). The free radical theory of aging is based on "rate of living hypothesis" which postulates species with higher metabolic rates age faster and has a shorter lifespan in comparison with species with a lower metabolic rate (Harman 1955). The lower metabolic rate extended lifespan in free-living nematode model *Caenorhabditis elegans* and fruit fly *Drosophila melanogaster* (Kenyon 2010). Similarly, a reduction in glucose levels in nutrient medium extended lifespan in *Saccharomyces cerevisiae* due to calorie restriction and lower metabolic rate (Kenyon 2010).

Another theory explaining the aging phenomenon is a theory of telomere shortening (Kirkwood 2005). Telomeres are repetitive short sequences of DNA located at the ends of eukaryotic chromosomes, which protects the chromosomes from degradation, fusion, and recombination (Kirkwood 2005). The DNA sequences at the telomeric end of each chromosome are not replicated during cell division in somatic cells (Kirkwood 2005). The reduction in the telomeric repeat is closely linked to cellular senescence and aging (Kirkwood 2005). The evidence demonstrating inhibition of cellular senescence by over-expression of telomerase enzyme restricting telomere shortening in cultured human cells supports the role of telomere shortening in aging (Kirkwood 2005).

The evolutionary theories argue that aging results from a decline in the force of natural selection (Kirkwood 2005), since evolution maximizes the reproductive fitness of an individual (Fischer et al. 2014). Hence, longevity is a trait to be selected only if it is beneficial for fitness. Therefore, lifespan may have a large degree of plasticity within an individual species and among species. Furthermore, the mutation accumulation theory of aging proposes accumulation of late-acting mutations which lead to age-related pathologies and death (Kirkwood 2005).

The concept of evolutionary trade-off is an integral part of disposable soma theory and antagonistic pleiotropy theory (Kirkwood 2005). The disposable soma theory describes the concept of lifespan but doesn't explain why we age, whereas the antagonistic pleiotropy theory suggests that the selection of some genes may be beneficial for early stages of life, but has detrimental effects with progression in age, thereby leading to senescence (Kirkwood 2005). The interaction between neuroendocrine and immune system plays a significant role in regulating lifespan (Rattan 2006). The interaction of neuropeptides, cytokines, and hormones can mediate pleiotropic changes regulating lifespan in an organism (Rattan 2006).

The gene regulation theory of aging postulates that changes in gene expression promote aging in living organisms. Earlier marking similarity in genetic and the biochemical pathways, regulating aging in organisms such as yeasts, flies, and mice hints at aging one of the programmed phenomenons in higher eukaryotes (Longo et al. 2015). The study of the effect of gene mutation in model organisms has demonstrated genetic interplay to regulate aging across various organisms (Kenyon et al. 1993; Longo et al. 2015). In fact, the very first mutation in *age-1* gene extended lifespan (65% increment in mean lifespan and 110% increase in maximum lifespan) in model organism *C. elegans* (Friedman and Johnson 1988). Since then, many lifespan-extending mutations have been discovered and most of them have been identified as homolog of majorly conserved cellular signaling pathways like insulin/IGF signaling pathway (insulin-like growth factor), *sir-2.1*, *skn-1*/*nrf-2* pathway, TOR signaling, and mitochondrial signaling (Baumeister et al. 2006).

These theories broaden our understanding of the aging process, and integration of these theories in various model systems can account for the phenomenon of aging. Altogether, these theories suggest that the pleiotropic changes with aging may result from one or more primary alterations that affect many downstream processes in living organisms.

2.3 *C. elegans* as a Model for Antiaging Studies

The gerontological research employs a range of model organisms such as *Escherichia coli*, *Saccharomyces cerevisiae*, *C. elegans*, *D. melanogaster*, and *Mus musculus* for studying the aging phenomenon (Longo et al. 2015). But free-living soil nematode has been widely exploited in gerontological studies owing to several advantageous features, viz. easy and low-cost maintenance and short lifespan in comparison with other model systems (Kenyon 2010). In the year 1965, Sydney Brenner introduced this free-living soil nematode as a model system (Brenner 1974). The tiny size of 1 mm, short life cycle, genetic heritability, and transparent body gives ease of access to study every single cell during development (Brenner 1974). The short lifespan of maximum 30 days in wild-type *C. elegans* (Fig. 2.3) makes this tiny worm an excellent model for studying aging (Kenyon 2010).

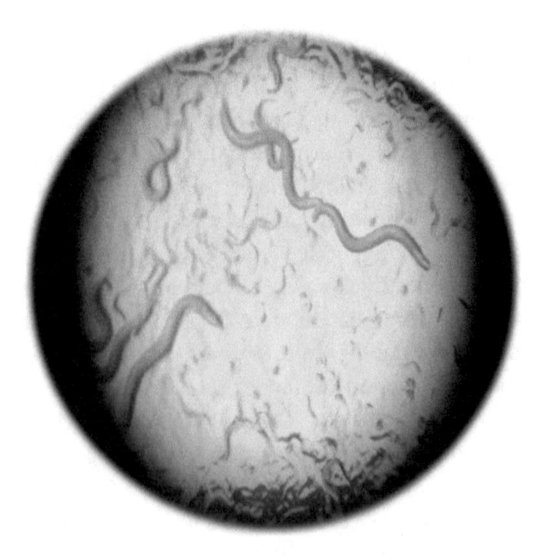

Fig. 2.3 Microscopic view of different stages of *C. elegans*

The generation time is about three days, which ranges from egg stage to adult stage of the worm. The cell lineage is completely known (Kenyon 2010). The occurrence of the male is very low, ranging from 0.01 to 0.1%, and hermaphrodites are more abundant and self-reproductive gender having 939 somatic cells in comparison with 1031 somatic cells in the males (Brenner 1974). The worm has 302 neurons, muscle cells, a reproductive tract, and an intestine (Kenyon 2010). In the year 1998, the complete genome was sequenced, and Sydney Brenner, H. Robert Horvitz, and John E. Sulston were jointly awarded the Nobel Prize in 2002 for their discoveries concerning genetic regulation of organ development and programmed cell death in *C. elegans*. The worm can be maintained on nematode growth medium agar Petri plates on *E. coli* lawn which is used as a food source or in axenic (without bacteria as a food source) or dead or alive *E. coli* containing liquid culture (Brenner 1974). The worms are grown at the ambient temperature ranging from 15 to 25 °C (Brenner 1974). The life cycle of the worm is temperature dependent; it is approximately 3 days at the temperature of 20 °C (Brenner 1974). There are four larval stages of the worm (L1–L4), and worms demonstrate developmentally arrested at L2 stage under adverse environmental conditions (Fig. 2.4). The worm at dauer stage can survive up to two months, and when food is available, dauer regains normal development (Brenner 1974). The worms are easy to maintain and can be frozen at −80 °C for long-term storage (Brenner 1974).

There are five pairs (I, II, III, IV, and V) of autosomal chromosomes and one pair of sex chromosome (Brenner 1974). The hermaphrodite can fertilize itself without mating (Brenner 1974). The short life cycle, ≈300 progeny, and ease of maintenance allow many assays to be carried out at once, which gives this model edge over in vitro and other cellular model systems (Brenner 1974). The 60–80% of genomic homology with humans, availability of the gene knockout (KO) mutant libraries, established genetic methodologies (forward and reverse), and well-established RNA interference (RNAi) mechanism provide a range of alternatives to manipulate and study *C. elegans* at the molecular level and elucidation of the associated genetic pathways (Tissenbaum 2015).

The growing evidence suggesting the role of plant-based molecule in delaying aging and age-related pathologies has given momentum to aging research. The advancement in the field of gerontology has proposed *C. elegans* as an excellent model for studying pharmacological manipulations. The discovery of multiple cellular pathways and genes regulating lifespan has led to the discovery of elicitors of longevity (Kenyon 2010). The previous studies report a range of plant molecules which promote healthy lifespan in *C. elegans* (Pant and Pandey 2015).

The higher homology of this nematode model with mammals extends the pharmacological application of discovered molecule in humans too (Garigan et al. 2002). These antiaging plant-based molecule basically found to modulate signaling pathways regulating aging (Pant and Pandey 2015). The health benefits of fruits, vegetable, and herbs can be attributed to this plant secondary metabolite. The *Ginkgo biloba,* apples, onions, red wine, and tea extract are rich in flavonoids like quercetin (Pant and Pandey 2015). Quercetin exposure prolongs lifespan in *C. elegans* by modulating intracellular ROS level, lipofuscin aggregation, and stress response by translocating DAF-*16* into the nucleus (Kampkötter et al. 2007). The green tea and tea components like catechins, epigallocatechin-3-gallate, and theaflavins are also found to promote longevity by modulating the stress response and expression of *daf-16* and its transcriptional target *sod-3* in *C. elegans* (Brown et al. 2006; Abbas and Wink 2009; Saul et al. 2009). Another flavonol, myricetin, also extended lifespan in *C. elegans* in *daf-16*-dependent manner (Pant and Pandey 2015). Furthermore, resveratrol was found to promote longevity and stress resistance in yeast *S. cerevisiae*, *C. elegans,* and *D. melanogaster* (Longo et al. 2015).

The commonly used spice curcumin promoted lifespan in *C. elegans* through protein homeostasis modulation (Liao et al. 2011). It was found to modulate the function of gerentogenes

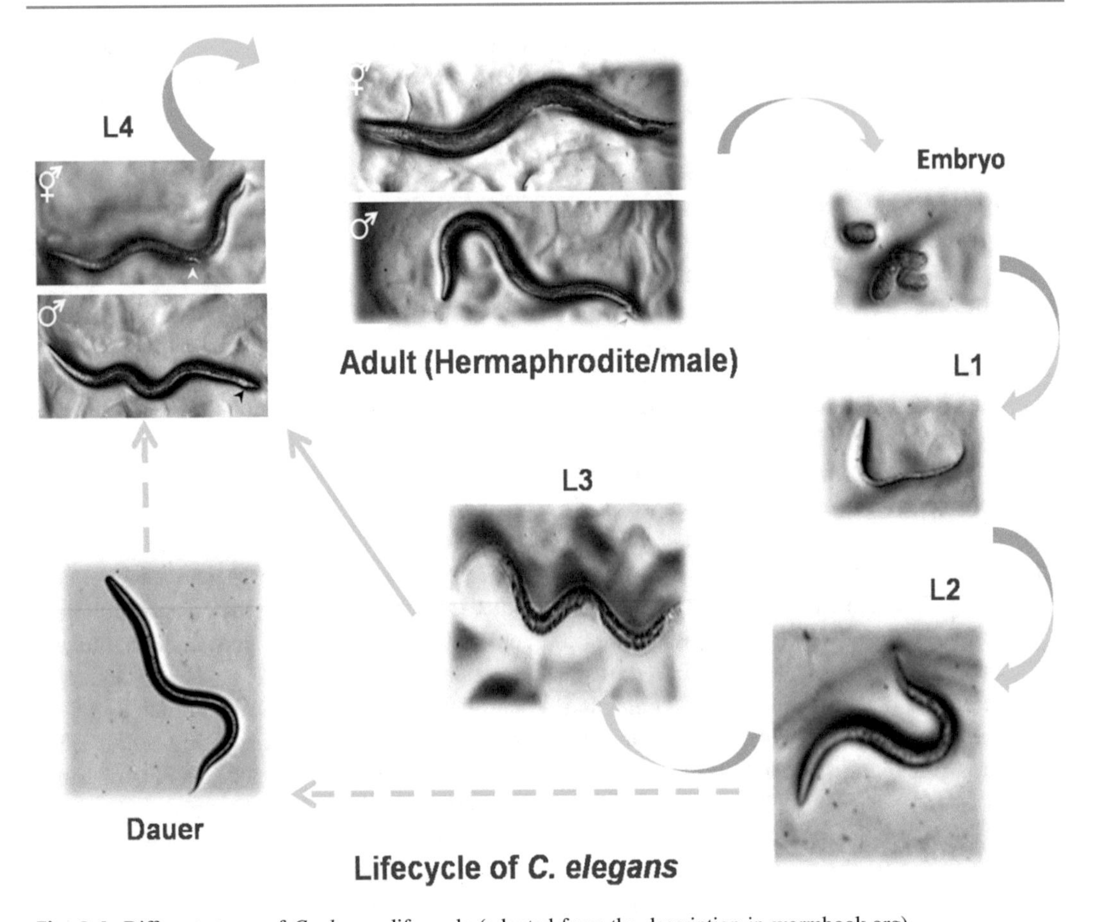

Fig. 2.4 Different stages of *C. elegans* life cycle (adapted from the description in wormbook.org)

such as *osr-1*, *sek-1*, *mek-1*, *skn-1*, *unc-43*, *sir-2.1*, and *age-1* mediating longevity in *C. elegans* (Kampkötter et al. 2007; Powolny et al. 2011; Pant and Pandey 2015). This bioactive phytomolecule majorly found as the component of *Ocimum,* and spice β-caryophyllene was able to modulate lifespan and stress level in *C. elegans*. It acts as the dietary restriction mimetic as it interacts with genetic elicitors of dietary restriction like *sir-2.1* and *skn-1* (Pant et al. 2014).

Additionally, another neuromodulatory compound reserpine exposure from embryo to young adult stage extended lifespan in *C. elegans* significantly (Srivastava et al. 2008). It is the major component of medicinal plant *Rauvolfia serpentina* and an FDA-approved drug for hypertension; it can be evaluated for lifespan extension in higher-model organisms like mice and finally humans. Harmane was able to modulate immune response, thereby prolonging lifespan in *C. elegans* (Jakobsen et al. 2013). The organosulfides stimulate phase II detoxification and cellular defence system ultimately extending lifespan in *C. elegans* (Powolny et al. 2011).

2.4 Medicinal Plants and Aging

The medicinal and aromatic plants (MAPs) are a rich repository of enormous bioactive phytomolecules present in the form of secondary metabolites. These plants have been exploited by the traditional system of medicine for treating severe ailments. The secondary metabolites derived from these medicinal plants have been utilized for decades in the form of drugs, antioxidants, flavors, and fragrances by more than 80% of the world population (Raskin et al. 2002).

The use of plants as medicine dates is as old as 4000–5000 BC, and Chinese used the natural herbal preparations as medicine for the first time. The beneficial effects of medicinal plants are well known owing to their natural origin and safer action (Lewis and Elvin-Lewis 1995). There are more than 3000 herbal medicines floating around the world, and plant-based therapeutics is gaining popularity around the world (Lewis and Elvin-Lewis 1995). The herbal medicines not only prevent disease, but also maintain the quality of life with reduced healthcare cost. The traditional system of Indian medicine "Ayurveda" is based on medicines of natural origin. The herbal formulations are often used for treating various age-related disorders like cardiovascular, cancer, arthritis, and neurodegenerative disorders (Lewis and Elvin-Lewis 1995; Kennedy and Wightman 2011). Many popular drugs are derived from plants, like morphine from *Papaver somniferum*, Ashwagandha from *Withania somnifera*, Ephedrine from *Ephedra vulgaris*, Atropine from *Atropa belladonna*, and Reserpine from *Rauwolfia serpentina*. These plants are rich in secondary metabolites and essential oils (which are potential sources of drugs) which are of high medicinal value. The important advantages claimed for therapeutic uses of medicinal plants in various ailments are their safety besides being economical, effective, and their easy availability (Lewis and Elvin-Lewis 1995). In India, the recent times have witnessed a resurgence of herbal medicine owing to hazardous side effects of synthetic drugs and least side effects of naturally derived phytomedicine (Kennedy and Wightman 2011). Because of these advantages, the medicinal plants have been widely used by the traditional medical practitioners in their day-to-day practice. According to a survey (1993) of World Health Organization (WHO), the practitioners of the traditional system of medicine treat about 80% of patients in India, 85% in Burma, and 90% in Bangladesh (Kennedy and Wightman 2011). The world market for herbal medicine is budding at an impeccable rate of 10–15% which is expected to grow at a higher rate by 2050 (Craig 1999; Pant and Pandey 2015).

The previous decade embarks the major discoveries unravelling the role of plant-based formulations and phytomolecules in defying aging and age-related pathologies (Longo et al. 2015; Pant and Pandey 2015). The medicinal herbs elicit beneficial effects due to the presence of a variety of phytomolecules such as flavonoids, terpenoids, and alkaloids (Pant and Pandey 2015). The phytomolecules are secondary metabolites which are regarded as biochemical "side tracks" with no role in plant's primary metabolism but have emerged as potential therapeutic agents (Craig 1999; Dillard and German 2000). However, the flavonoids, terpenoids, and alkaloids play a significant role in plant defense system (Kennedy and Wightman 2011). These molecules have antioxidant and anti-inflammatory properties which make them a good option for targeting aging and age-related pathologies (Powolny et al. 2011).

Various researchers have demonstrated the lifespan-prolonging and stress modulatory effects of natural molecules derived from medicinal plants (Kampkötter et al. 2007; Wilson et al. 2006; Zhang et al. 2009; Powolny et al. 2011). These studies demonstrate that phytomolecules modulate the stress level and promote longevity in invertebrate models (Table 2.1). The flavonoids, alkaloids, anthocyanin, glycosides, and proanthocyanidin oligomers have demonstrated stress modulatory and lifespan-prolonging effects (Longo et al. 2015).

In recent years, several breakthroughs have been witnessed in phytomolecule research where antiaging and stress modulatory activities of various plants like *Ginkgo biloba*, *Vitis vinifera*, *Eleutherococcus senticosus*, *Rhodiola rosea*, *Cinnamomum cassia*, and *Panax ginseng* (Rai et al. 2003; Liu et al. 2012; Wilson et al. 2006; Kampkötter et al. 2007; Zhang et al. 2009; Yu et al. 2010) have been unravelled. The least side effects of phytomolecules like resveratrol, quercetin, diallyltrisulfide, curcumin, β-caryophyllene, and reserpine which counteract aging make them highly sought after for treating age-related pathologies (Pant et al. 2014). The prime goal of medicinal plant research is to screen out the best herbal formulation/active

Table 2.1 Plant extract/phytomolecules increase lifespan in *C. elegans*

S. No.	Phytomolecules	Plant source		Antiaging activities and mode of action	References
1.	Vitamin E	Olive oil, coconut oil, sunflower oil, broccoli, kale, spinach, avocado, almonds, peanuts	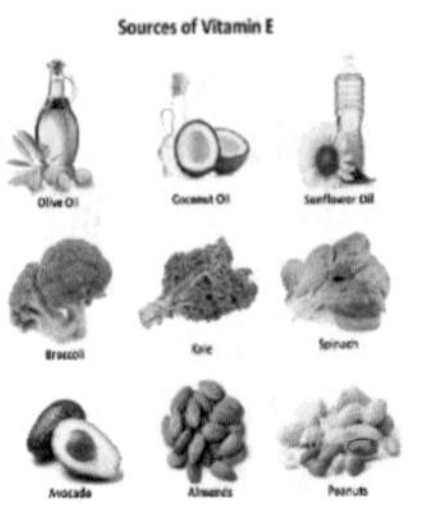	Antioxidant	Harrington and Harley (1988)
2.	EGb 761	*Ginkgo biloba*		Stress resistance	Wu et al. (2002)
3.	Resveratrol	Grapes		Stress resistance, autophagy	Bass et al. (2007), Wood et al. (2004)
4.	Blueberry extract	Blueberries		Osmotic pathway	Peng et al. (2012), Wilson et al. (2006)
5.	Lipoic acid	Tomato, broccoli, spinach		Stress resistance	Brown et al. (2006)
6.	Reserpine	*Rauwolfia serpentina*		Serotonin signaling	Srivastava et al. (2008)

(continued)

Table 2.1 (continued)

S. No.	Phytomolecules	Plant source		Antiaging activities and mode of action	References
7.	Quercetin	Apple, strawberries, onion		DAF-16 dependent	Pietsch et al. (2011)
8.	Catechin	*Camellia sinensis*		DAF-2, MEV-1, and NHR-8 dependent	Saul et al. (2009)
9.	Cinnamomum cassia Bark	*Cinnamomum cassia*		DAF-16, MEV-1, and SER-1 dependent	Yu et al. (2010)
10.	Curcumin	*Curcuma longa*		Protein homeostasis	Liao et al. (2011)
11.	Diallyltrisulfide	Garlic, carrots, and broccoli		SKN-1 dependent	Powolny et al. (2011)
12.	Royal jelly	Royal jelly and honey		DAF-2/DAF-16 dependent	Honda et al. (2011)
13.	Acacetin 7-*O*-α-l-rhamnopyranosyl (1–2) β-D-xylopyranoside	*Premna integrifolia*		SIR-2.1, SAMS-1, dietary restriction, and stress response	Asthana et al. (2015a, b)

(continued)

Table 2.1 (continued)

S. No.	Phytomolecules	Plant source		Antiaging activities and mode of action	References
14.	β-caryophyllene	*Ocimum* and clove		SKN-1, SIR-2.1, DAF-16, and xenobiotic stress response pathways	Pant et al. (2014)
15.	4-HEG	*Premna integrifolia*		Stress response	Shukla et al. (2012)

phytomolecule for modulating aging and preventing age-related pathologies. The medicinal plant extracts and phytomolecules have been reported to possess antioxidant and stress modulatory potential which makes them an ideal candidate for antiaging research. As progression in age and age-related disorders is associated with enhanced oxidative stress and susceptibility to environmental stress, therefore, the medicinal plants having high antioxidant and stress modulatory potential can serve as an ideal candidate for combating aging-related decrepitude.

The reference to the use of plants as a medicine dates back to 3000–4000 BC as per writings in *Rigveda*. Later on, medicinal properties of many plants and different combinations of herbal formulations were studied and described by ancient practitioners of Ayurvedic medicine which forms the basic foundation of Indian system of medicine. The Ayurvedic literature dates back to previous centuries full of text describing the therapeutic application of medicinal plants such as Ashwagandha (*Withania somnifera*), Sarpagandha (*Rauwolfia serpentina*), Brahmi (*Bacopa monnieri*), Clove (*Syzygium aromaticum*), Tulsi (*Ocimum sanctum*), Haldi (*Curcuma longa*), Adrak or Ginger (*Zingiber officinale*), Garlic (*Allium sativum*), Sadabahar (*Catharanthus roseus*), and many more (Gurib-Fakim 2006). Therefore, there is a need for exploring the detailed mode of action of these rich herbs for the benefit of mankind. In Ayurveda, Tulsi which is scientifically known as *Ocimum sanctum* L. has been well documented for its various therapeutic potential and described as "*Dashemani Shwasaharni*" (antiasthmatic) and antiseptic drugs (Kaphaghna) (Khan and Balick 2001). Although *Ocimum* spp. have been widely exploited for its medicinal properties by traditional practitioners of Indian and Chinese medicine, wide scientific validation is still not much available. In recent years, many scientific studies have been performed to elucidate the potential role of essential oil and phytomolecules isolated from this therapeutically important medicinal plant (Pant and Pandey 2015). Traditionally, different parts of this plant have been used for treating various ailments, and in recent years, various molecules isolated from genus *Ocimum* have been scientifically evaluated for their therapeutic potential (Chopra and Doiphode 2002). The pharmacological importance of *Ocimum* spp. has been reported and described by some researchers recently (Pandey et al. 2013). The present chapter highlights the antiaging properties of *Ocimum* spp. and its potential as a therapeutic agent in treating various age-related diseases.

2.5 *Ocimum* spp

The botanical name *Ocimum* is derived from the Greek meaning "to be fragrant." In the 1600 s, the English used basil as a food flavoring and insecticide (Gille et al. 2007). Different parts of this plant are used in Ayurveda and Siddha systems of medicine for prevention and cure of many illnesses (Govindarajan et al. 2005). This plant has been part of various cultures (Fig. 2.5). The Italians used basil as the sign of love (Gille et al. 2007). Hindus believed that being buried with a leaf of basil will get them salvation (Khosla 1995). Basil is believed to be sacred to the gods Krishna and Vishnu in India. Americans have been growing basil for over 200 years; they used to air-dry the plant and preserve in layers of salt in earthen pots (Joshi 2017). The plants of this genus are generally erect, fragrant, branched, and about 20–50 cm in height (Baseer and Jain 2016). The leaves are aromatic, simple, opposite, elliptic, and acute with entire or sub-serrate or dentate margins, growing up to 5 cm long (Baseer and Jain 2016). The flowers are small having purple to reddish color, present in small compact clusters on cylindrical spikes (Baseer and Jain 2016). The flowers are small in size and rarely 5 mm long (Baseer and Jain 2016). The traditional system of Indian medicine utilizes different parts of this plant viz. leaves, stem, flower, root, seeds, and even whole plant for treating severe ailments like bronchitis, bronchial asthma, malaria, diarrhea, dysentery, skin diseases, arthritis, painful eye diseases, chronic fever, and insect bite (Prakash and Gupta 2005). Apart from treating common ailments, it is also suggested to possess anticancer, antidiabetic, antifungal, cardioprotective, antifertility, antimicrobial, antispasmodic, analgesic, adaptogenic, and diaphoretic potential (Khosla 1995). The widely studied variety of *Ocimum* sps. viz. *Ocimum sanctum* and *Ocimum basilicum* has been suggested to possess therapeutic potential and their active constituent like eugenol, oleanolic acid, ursolic acid, rosmarinic acid, limonene, carvacrol, linalool, and β-caryophyllene that found to be majorly responsible for the therapeutic potential of *Ocimum sps* (Gille et al. 2007; Baseer and Jain 2016). Despite its wide exploitation by practitioners of traditional medicine owing to its high therapeutic value and wide occurrence in India, scientific validation of this medicinal herb is still very limited. In order to establish its potential as a therapeutic solution to various ailments in modern medicine, the world community of scientist has extensively studied the pharmacological properties of various extracts and molecules isolated from *Ocimum sps.* (Khosla 1995; Bilal et al. 2012; Pingale et al. 2012). The outcome of these studies has established its therapeutic potential for curing various ailments.

Fig. 2.5 *Ocimum sanctum* "Queen of herbs"

2.6 Phytoconstituents of *Ocimum* spp

The aromatic odor of *Ocimum* sps. (*O. sanctum*) is mainly due to the presence of essential oil that is mainly concentrated in its leaf (Prakash and Gupta 2005). The essential oil mainly contains phenols, terpenes, and aldehydes as its major constituent (Baseer and Jain 2016). The essential oil isolated from *Ocimum* spp. mainly comprises of compounds like myrcene, linalool, eugenol, β-caryophyllene, iso-eugenol, geraniol, methyl eugenol, α-humulene, toluene, camphene, sabinene, dimethyl benzene, α-thujene, octane, ethyl benzene, limonene, p-cymene, terpinolene, allo-cimene, butyl-benzene, α-cubebene, nonane, α-pinene, pinene, elemene, lactate, α-guaiene, α-amorphene, humulene, calamine, nerolidol, iedol and elemol, α-terpineol, α-murolene, cadinene isoborneol, borneol, α-selinene, carvacrol, and selinene (Baseer and Jain 2016). The major class of compound that is present in essential oil of *Ocimum* sps. is terpenes (Gille et al. 2007).

2.7 *Ocimum* sps: The Elixir of Life

The medicinal and aromatic plant *O. sanctum* is commonly known as "Queen of herbs" and locally as "Tulsi" in India (Joshi 2017). *O. sanctum* (Fig. 2.1) is the most prominent species of the *Ocimum* genera (Kumar et al. 2011). Tulsi is also known as "the elixir of life" since it promotes longevity. Different parts of this plant have been used by traditional practitioner of Ayurveda and Siddha systems of medicine for treating various ailments like influenza, headache, bronchitis, asthma, common cold, cough, colic pain, earache, fever, sore throat, malarial fever, hepatic diseases, flatulence, diarrhea, migraine headaches, fatigue, as an antidote for snake bite and scorpion sting, skin diseases, wound, arthritis, night blindness, and insomnia (Prakash and Gupta 2005; Kumar et al. 2011; Bilal et al. 2012; Pingale et al. 2012; Monga et al. 2017). The consumption of few leaves every day is considered good for nerves and memory and can cure ulcers of mouth (Khosla 1995). *Ocimum* sps. are known to demonstrate immunomodulatory effects, and some reports suggest that it boosts up the immune system functioning (Mondal et al. 2011). The extract of *Ocimum* leaves protects from all sorts of infection from bacteria, fungi, protozoa, and viruses (Chiang et al. 2005; Prakash and Gupta 2005). The fresh leaves and stem of *Ocimum sps*. extract possess bioactive compounds like phenols (antioxidants) such as apigenin, cirsilineol, cirsimaritin, rosmarinic acid, isothymusin, and eugenol. The leaves of *O. sanctum* contain 0.7% volatile oil which constitutes approximately 71% eugenol and 20% methyl eugenol. The oil contains carvacrol and sesquiterpene β-caryophyllene as additional components. The aqueous extract of leaves from *O. sanctum* contains flavonoids like orientin and vicenin (Gille et al. 2007). The recent studies on elucidating the therapeutic potential of *Ocimum* sps. found that it inhibits HIV growth and proliferation of cancer cell (Rege and Chowdhary 2014). The researchers studied effect of *O. sanctum* extract exposure on mice having sarcoma-180 solid tumors and *Ocimum*-treated mice demonstrated a significant reduction in tumor cell size and extension in lifespan (Karthikeyan et al. 1999). Similar to these results, another research group administered leaf extract of *O. sanctum* orally (200 mg/kg, per organism) and observed significantly reduced tumor volume, elevated body weight, and survival rate in mice (Karthikeyan et al. 1999). In addition to that, recent study reports the protective effect of *Ocimum* sps. against radiation-induced DNA damage (Güez et al. 2017). Additionally, different extracts of *O. sanctum* (aqueous, alcoholic, chloroform extract) and essential oil were found to exhibit antibacterial activity against *E. coli*, *P. aeruginosa*, *S. typhimurium*, and *S. aureus* (Kumar et al. 2011). The *Ocimum* extract was found to be equally effective against pathogenic gram-positive and gram-negative bacteria (Joshi 2017). The *O. basilicum* was also found to possess potential antiaging property as an active formulation containing basil extract when applied topically (as the stable topical emulsion) delayed skin aging and reduced appearance of wrinkles (Jadoon et al. 2015). The active constituent of basil and their coordinating action imparts the antiaging properties to the cream formulation that rejuvenates skin (Raskin et al. 2002). The presence of high phenolic content and flavonoids makes basil extract effective against UVR-induced damage (Jadoon et al. 2015). The methanolic extract of *O. sanctum* demonstrated cardioprotective properties as its exposure reduced isoproterenol (ISP)-induced myocardial infarction (Kavitha et al. 2015). The pre-treatment with *Ocimum* extract significantly reduced the level of cardiac markers, phospholipases, and phospholipid content in ISP-treated rats (Kavitha et al. 2015). In addition to that, rats fed with *Ocimum* extract demonstrated reduced 5-lipoxygenase and cyclooxygenase-2 activities and leukotriene B4 and thromboxane B2 level in ISP-treated animals (Kavitha et al. 2015). The high phenolic content is demonstrated as the reason behind the cardioprotective effect of *Ocimum* extract in this study (Kavitha et al. 2015). The high phenolic contents and antioxidant potential reduced the oxidative stress and modulated the arachidonic acid pathway which in turn was found to promote survival in ISP-induced rats (Kavitha et al. 2015). In another study, exposure to *O. sanctum* extract for 21 days was found to reduce oxidative stress level in brain's frontal cortex and striatum in rat model (Bora et al. 2011). The *O. sanctum* extract administration reduced hepatic liver peroxidation induced by iron overload suggesting stress modulating potential of *Ocimum sps* (Muralikrishnan et al. 2012). These outcomes which support antioxidant potential of *Ocimum sps.* form the basis for the use of *O. sanctum* for clinical conditions involving elevated oxidative stress. The formulation containing *O. sanctum* improved learning and memory in normal and memory-impaired rats suggesting its potential role in alleviating age-related cognitive decline (Malve et al. 2014). Another species of *Ocimum,* i.e., *O. gratissimum,* also demonstrated cardioprotective role in one study where it was able to reduce hypercholesterolemia and hypertriglyceridemia induced by fat and cholesterol-rich diet (Chao et al. 2016). The *O. gratissimum* exposure reduced total cholesterol, triglycerides, low-density lipoprotein, and improved high-density lipoprotein levels suggesting potential cardioprotective activity (Chao et al. 2016). In another study, *O. basilicum* extract demonstrated antioxidant activity as it was able to effectively revert the effect of highly oxidizing hydrogen peroxide (Muralikrishnan et al. 2012; Asthana et al. 2015a, b). The antioxidant effect demonstrated by basil extract was an outcome of high polyphenols and flavonoids like rosmarinic acid and eugenol, and alkaloid like β-caryophyllene which are well known as antioxidant compounds (Asthana 2015c). The *Ocimum* sps. was found to reduce oxidative stress, improve survival, and reduce age-related decline in cognition as per these studies which induced age-related stress phenotype in mice and rat model. The stress modulatory and enhanced survival in organism demonstrated by *Ocimum* sps. can be due to the presence of high level of bioactive compounds like eugenol, rosmarinic acid, β-caryophyllene, ursolic acid, and oleanolic acid. Recent studies demonstrating antiaging effect of these molecules in model organisms support the potential role of *Ocimum sps.* rich in these molecules as an anti-aging medicinal herb (Asthana et al. 2015a, b). One of the active ingredient oleanolic acids which is already known to have many pharmacological activities was evaluated for its lifespan and stress modulatory potential employing *C. elegans* model system (Zhang et al. 2015). The outcome of the study suggests that oleanolic acid promotes lifespan and stress resistance in aging worms (Zhang et al. 2015). The oleanolic acid was found to promote life in DAF-16/FOXO (forkhead transcription factor in humans)-dependent manner (Zhang et al. 2015). The oleanolic acid was able to modulate insulin signaling pathway which is also found to regulate metabolism and lifespan in humans (Zhang et al. 2015). The oleanolic acid exposure enhanced expression of stress response gene like superoxide dismutase and heat shock proteins in worms suggesting stress modulatory effects of this active ingredient of basil (Zhang et al. 2015). In another study, one of the major components of *Ocimum* sps. ursolic acid improved age-related decline in skeletal muscle cell performance and cellular energy status (Bahrami and Bakhtiari 2016). In addition to this study, another research group found ursolic acid supplementation elevated antiaging biomarkers like (SIRT1 and SIRT6) and PGC-1β in hypothalamus which is known to regulate aging and age-related mitochondrial disorders (Bahrami and Bakhtiari 2016; Liu 1995). In addition to this study, ursolic acid was also able to prolong survival by 30% and reduce stress in *C. elegans* by modulating JNK-1 expression and its downstream targets (Negi et al. 2016, 2017). Another bioactive component of *Ocimum* sps. is rosmarinic acid which was also found to promote lifespan and alleviate age-related oxidative stress utilizing *C. elegans* as model for aging (Wang et al. 2012). The antiaging potential demonstrated by rosmarinic acid was an outcome of alteration in lifespan-regulated signaling network regulated by genes like *osr-1, sek-1, sir-2.1, and daf-16* which are also conserved in humans (Wang et al. 2012). Another important component of this medicinal herb is β-caryophyllene, which is considered as highly antioxidant phytomolecule (Pant et al. 2014); β-caryophyllene exposure was able to reduce age-related decline in *C. elegans* (Pant et al. 2014). The β-caryophyllene pre-treated animals demonstrated enhanced lifespan by over 22% and reduced intracellular free radical levels (Pant et al. 2014). The β-caryophyllene-treated animals demonstrated reduced oxidative stress level and prolonged survival (Pant et al. 2014). The β-caryophyllene-mediated antiaging effects are found to be dependent on SIR-*2.1*, SKN-*1*, and DAF-*16* as β-caryophyllene pre-treated animals demonstrated increased expression of these gerontogenes and their downstream targets, which regulate major aging pathways (Pant et al. 2014). Altogether these recent studies suggest that *Ocimum* sps. possess antiaging and stress modulatory potential which can be used for the benefit of mankind. These studies are able to scientifically evaluate the antiaging potential of this traditionally used herb which is cited as longevity elixir due to the presence of many bioactive compounds which possess high antioxidant potential and stress modulatory properties. The presence of active components like ursolic acid, rosmarinic acid, eugenol, and β-caryophyllene which have been individually evaluated for their lifespan and age-related stress modulatory effects and found to possess antiaging potential. The future investigations highlighting the clinical evidence supporting of *Ocimum* sps. as an antiaging and stress modulatory herb will be helpful in designing herb-based formulations for treating age-related pathologies.

References

Abbas S, Wink M (2009) Epigallocatechin gallate from green tea (*Camellia sinensis*) increases lifespan and stress resistance in *Caenorhabditis elegans*. Planta Med 75(3):216–221

Asthana J, Pant A, Yadav D, Lal R, Gupta M et al (2015a) *Ocimum basilicum* (L.) and *Premna integrifolia* (L.) modulate stress response and lifespan in *Caenorhabditis elegans*. Indust Crops Prod 76(8):1086–1093

Asthana J, Yadav D, Pant A, Yadav A, Gupta M et al (2015b) Acacetin 7-O-α-l-rhamnopyranosyl (1–2) β-D-xylopyranoside elicits life-span extension and stress resistance in *Caenorhabditis elegans*. J Gerontol Ser A: Biol Sci Med Sci 71(9):1160–1168

Asthana J, Yadav AK, Pant A, Pandey S, Gupta MM et al (2015c) Specioside ameliorates oxidative stress and promotes longevity in Caenorhabditis elegans. Comp Biochem Physiol C Toxicol Pharmacol 169:25–34

Bahrami SA, Bakhtiari N (2016) Ursolic acid regulates aging process through enhancing of metabolic sensor proteins level. Biomed Pharmacother 82(8):8–14

Baseer M, Jain K (2016) Review of botany, phytochemistry, pharmacology, contemporary applications and toxicology of *Ocimum sanctum*. Int J Pharm Life Sci 7 (2):4918–4929

Bass TM, Weinkove D, Houthoofd K, Gems D, Partridge L (2007) Effects of resveratrol on lifespan in *Drosophila melanogaster* and *Caenorhabditis elegans*. Mech Age Dev 128(10):546–552

Baumeister R, Schaffitzel E, Hertweck M (2006) Endocrine signaling in *Caenorhabditis elegans* controls stress response and longevity. J Endocrinol 190(2): 191–202

Bilal A, Jahan N, Ahmed A, Bilal SN, Habib S et al (2012) Phytochemical and pharmacological studies on *Ocimum basilicum* Linn-a review. Int J Curr Res Rev 4(23):73–83

Blagosklonny MV (2009) Validation of anti-aging drugs by treating age-related diseases. Aging 1(3):281–288

Bora KS, Arora S, Shri R (2011) Role of *Ocimum basilicum* L. in prevention of ischemia and reperfusion-induced cerebral damage, and motor dysfunctions in mice brain. J Ethnopharmacol 137(3): 1360–1365

Brenner S (1974) The genetics of *Caenorhabditis elegans*. Genetics 77(1):71–94

Brown MK, Evans JL, Luo Y (2006) Beneficial effects of natural antioxidants EGCG and α-lipoic acid on life span and age-dependent behavioral declines in *Caenorhabditis elegans*. Pharmacol Biochem Behav 85(3):620–628

Chao PY, Lin JA, Ting WJ, Lee HH, Hsieh K et al (2016) *Ocimum gratissmum* aqueous extract reduces plasma lipid in hypercholesterol-fed hamsters. Int J Med Sci 13(11):819–824

Chiang LC, Ng LT, Cheng PW, Chiang W, Lin CC (2005) Antiviral activities of extracts and selected pure constituents of *Ocimum basilicum*. Clin Exp Pharmac Physiol 32(10):811–816

Chopra A, Doiphode VV (2002) Ayurvedic medicine: core concept, therapeutic principles, and current relevance. Med Clin North Am 86(1):75–89

Craig WJ (1999) Health-promoting properties of common herbs. Am J Clin Nutr 70(3):491–499

da Costa JP, Vitorino R, Silva GM, Vogel C, Duarte AC et al (2016) A synopsis on aging—theories, mechanisms and future prospects. Age Res Rev 29:90–112

Dillard CJ, German JB (2000) Phytochemicals: nutraceuticals and human health. J Sci Food Agri 80(12):1744–1756

Fischer B, van Doorn GS, Dieckmann U, Taborsky B (2014) The evolution of age-dependent plasticity. Am Nat 183(1):108–125

Friedman DB, Johnson TE (1988) A mutation in the *age-1* gene in *Caenorhabditis elegans* lengthens life and reduces hermaphrodite fertility. Genetics 118(1): 75–86

Garigan D, Hsu AL, Fraser AG, Kamath RS, Ahringer J et al (2002) Genetic analysis of tissue aging in *Caenorhabditis elegans*: a role for heat-shock factor and bacterial proliferation. Genetics 161(3):1101–1112

Gille E, Danila D, Stanescu U, Hancianu M (2007) The phytochemical evaluation of some extracts of *Ocimum sp*. Planta Med 73(09):374

Govindarajan R, Vijayakumar M, Pushpangadan P (2005) Antioxidant approach to disease management and the role of ‘Rasayana’ herbs of Ayurveda. J Ethnopharmacol 99(2):165–178

Güez CM, Souza ROD, Fischer P, Leão MFDM, Duarte JA et al (2017) Evaluation of basil extract (*Ocimum basilicum* L.) on oxidative, anti-genotoxic and anti-inflammatory effects in human leukocytes cell cultures exposed to challenging agents. Braz J Pharmaceut Sci 53(1):1–12

Gurib-Fakim A (2006) Medicinal plants: traditions of yesterday and drugs of tomorrow. Mol Aspects Med 27(1):1–93

Harman D (1955) Aging: a theory based on free radical and radiation chemistry. Sci Aging Knowl Environ 2002(37):298–300

Harrington LA, Harley CB (1988) Effect of vitamin E on lifespan and reproduction in *Caenorhabditis elegans*. Mech Age Dev 43(1):71–78

Honda Y, Fujita Y, Maruyama H, Araki Y, Ichihara K et al (2011) Lifespan-extending effects of royal jelly and its related substances on the nematode *Caenorhabditis elegans*. PLoS ONE 6(8):e23527

Jadoon S, Karim S, Asad MHHB, Akram MR, Kalsoom Khan A et al (2015) Anti-aging potential of phytoextract loaded-pharmaceutical creams for human skin cell longetivity. Oxid Med Cell Longetivity 3:1–12 2015

Jakobsen H, Bojer MS, Marinus MG, Xu T, Struve C et al (2013) The alkaloid compound harmane increases the lifespan of *Caenorhabditis elegans* during bacterial infection, by modulating the nematode’s innate immune response. PLoS ONE 8(3):e60519

Joshi RK (2017) Phytoconstituents, traditional, medicinal and bioactive uses of Tulsi (*Ocimum sanctum* Linn.): a review. J Pharmacognosy Phytochem 6(2):261–264

Kampkötter A, Nkwonkam CG, Zurawski RF, Timpel C, Chovolou Y et al (2007) Investigations of protective effects of the flavonoids quercetin and rutin on stress resistance in the model organism *Caenorhabditis elegans*. Toxicology 234(1):113–123

Karthikeyan K, Gunasekaran P, Ramamurthy N, Govindasamy S (1999) Anticancer activity of *Ocimum sanctum*. Pharmaceut Biol 37(4):285–290

Kavitha S, John F, Indira M (2015) Amelioration of inflammation by phenolic rich methanolic extract of *Ocimum sanctum* Linn. leaves in isoproterenol induced myocardial infarction. Indian J Exp Biol 53(10):632–640

Kennedy DO, Wightman EL (2011) Herbal extracts and phytochemicals: plant secondary metabolites and the enhancement of human brain function. Adv Nutr Int Rev J 2(1):32–50

Kenyon CJ (2010) The genetics of ageing. Nature 464 (7288):504–512

Kenyon C, Chang J, Gensch E, Rudner A, Tabtiang R (1993) A *C. elegans* mutant that lives twice as long as wild type. Nature 366(6454):461–464

Khan S, Balick MJ (2001) Therapeutic plants of Ayurveda: a review of selected clinical and other studies for 166 species. J Alt Comp Med 7(5):405–515

Khosla M (1995) Sacred tulsi (*Ocimum sanctum* L.) in traditional medicine and pharmacology. Anc Sci Life 15(1):53

Kirkwood TB (2005) Understanding the odd science of aging. Cell 120(4):437–447

Kumar V, Andola HC, Lohani H, Chauhan N (2011) Pharmacological review on *Ocimum* sanctum Linnaeus: a queen of herbs. J Pharm Res 4(2):366–368

Lewis WH, Elvin-Lewis MP (1995) Medicinal plants as sources of new therapeutics. Ann Mo Bot Gard 82 (1):16–24

Liao VHC, Yu CW, Chu YJ, Li WH, Hsieh YC et al (2011) Curcumin-mediated lifespan extension in *Caenorhabditis elegans*. Mech Age Dev 132(10):480–487

Liu J (1995) Pharmacology of oleanolic acid and ursolic acid. J Ethnopharmacol 49(2):57–68

Liu Z, Li X, Simoneau AR, Jafari M, Zi X (2012) *Rhodiola rosea* extracts and salidroside decrease the growth of bladder cancer cell lines via inhibition of the mTOR pathway and induction of autophagy. Mol Carc 51(3):257–267

Longo VD, Antebi A, Bartke A, Barzilai N, Brown-Borg HM et al (2015) Interventions to slow aging in humans: are we ready? Aging Cell 14(4):497–510

Malve HO, Raut SB, Marathe PA, Rege NN (2014) Effect of combination of *Phyllanthus emblica*, *Tinospora cordifolia*, and *Ocimum sanctum* on spatial learning and memory in rats. J Ayurveda Int Med 5(4):209–215

Mondal S, Varma S, Bamola VD, Naik SN, Mirdha BR et al (2011) Double-blinded randomized controlled trial for immunomodulatory effects of Tulsi (*Ocimum sanctum* Linn.) leaf extract on healthy volunteers. J Ethnopharmacol 136(3):452–456

Monga S, Dhanwal P, Kumar R, Kumar A, Chhokar V (2017) Pharmacological and physico-chemical properties of Tulsi (*Ocimum gratissimum* L.): An updated review. Pharma Innov 6(4):181–186

Muralikrishnan G, Pillai S, Shakeel F (2012) Protective effects of *Ocimum sanctum* on lipid peroxidation and antioxidant status in streptozocin-induced diabetic rats. Nat Prod Res 26(5):474–478

Negi H, Shukla A, Khan F, Pandey R (2016) 3β-Hydroxy-urs-12-en-28-oic acid prolongs lifespan in *C. elegans* by modulating JNK-1. Biochem Biophys Res Comm 480(4):539–543

Negi H, Saikia SK, Pandey R (2017) 3β-Hydroxy-urs-12-en-28-oic acid modulates dietary restriction mediated longevity and ameliorates toxic protein aggregation in *C. elegans*. J Gerontol Ser A: Biomed Sci Med Sci 72(12):1614–1619

Pandey R, Gupta S, Shukla V, Tandon S, Shukla V (2013) Antiaging, antistress and ROS scavenging activity of crude extract of *Ocimum sanctum* (L.) in *Caenorhabditis elegans* (Maupas, 1900). Indian J Exp Biol 51:515–521

Pant A, Pandey R (2015) Bioactive phytomolecules and aging in *Caenorhabditis elegans*. Healthy Aging Res 4(19):1–15

Pant A, Saikia SK, Shukla V, Asthana J, Akhoon BA, Pandey R (2014) Beta-caryophyllene modulates expression of stress response genes and mediates longevity in *Caenorhabditis elegans*. Exp Gerontol 57:81–95

Peng C, Zuo Y, Kwan KM, Liang Y, Ma KY et al (2012) Blueberry extract prolongs lifespan of Drosophila melanogaster. Exp Gerontol 47(2):170–178

Pietsch K, Saul N, Chakrabarti S, Stürzenbaum SR, Menzel R et al (2011) Hormetins, antioxidants and prooxidants: defining quercetin-, caffeic acid-and rosmarinic acid-mediated life extension in *C. elegans*. Biogerontology 12(4):329–347

Pingale SS, Firke NP, Markandetya A (2012) Therapeutic activities of *Ocimum* tenuiflorum accounted in last decade: a review. J Pharm Res 5(4):2215–2220

Powolny AA, Singh SV, Melov S, Hubbard A, Fisher AL (2011) The garlic constituent diallyl trisulfide increases the lifespan of *C. elegans* via *skn-1* activation. Exp Gerontol 46(6):441–452

Prakash P, Gupta N (2005) Therapeutic uses of *Ocimum sanctum* Linn (Tulsi) with a note on eugenol and its pharmacological actions: a short review. Indian J Physiol Pharmacol 49(2):659–666

Rai D, Bhatia G, Sen T, Palit G (2003) Anti-stress effects of *Ginkgo biloba* and *Panax ginseng*: a comparative study. J Pharmacol Sci 93(4):458–464

Raskin I, Ribnicky DM, Komarnytsky S, Ilic N, Poulev A et al (2002) Plants and human health in the twenty-first century. Trends Biotechnol 20(12):522–531

Rattan SI (2006) Theories of biological aging: genes, proteins, and free radicals. Free Rad Res 40(12):1230–1238

Rege A, Chowdhary AS (2014) Evaluation of *Ocimum sanctum* and *Tinospora cordifolia* as probable HIV protease inhibitors. Int J Pharmaceut Sci Re Res 25 (1):315–318

Saul N, Pietsch K, Menzel R, Stürzenbaum SR, Steinberg CE (2009) Catechin induced longevity in *C. elegans*: from key regulator genes to disposable soma. Mech Ageing Dev 130(8):477–486

Shukla V, Yadav D, Phulara SC, Gupta M, Saikia SK, Pandey R (2012) Longevity-promoting effects of 4-hydroxy-E-globularinin in *Caenorhabditis elegans*. Free Rad Biol Med 53(10):1848–1856

Srivastava D, Arya U, SoundaraRajan T, Dwivedi H, Kumar S, Subramaniam JR (2008) Reserpine can confer stress tolerance and lifespan extension in the nematode *C. elegans*. Biogerontology 9(5):309–316

Tissenbaum HA (2015) Using *C. elegans* for aging research. Inv Repro Dev 59(12):59–63

Wang F, Liu QD, Wang L, Zhang Q, Hua ZT (2012) The molecular mechanism of rosmarinic acid extending the lifespan of *Caenorhabditis elegans*. App Mech Mat 140(11):469–472

Wilson MA, Shukitt Hale B, Kalt W, Ingram DK, Joseph JA (2006) Blueberry polyphenols increase lifespan and thermotolerance in *Caenorhabditis elegans*. Aging Cell 5(1):59–68

Wood JG, Rogina B, Lavu S, Howitz K, Helfand SL (2004) Sirtuin activators mimic caloric restriction and delay ageing in metazoans. Nature 430(7000):686–689

Wu Z, Smith JV, Paramasivam V, Butko P, Khan I (2002) Ginkgo *biloba* extract EGb 761 increases stress resistance and extends life span of *Caenorhabditis elegans*. Cell Mol Biol 48(6):725–731

Yu YB, Dosanjh L, Lao L, Tan M, Shim BS (2010) *Cinnamomum cassia* bark in two herbal formulas increases life span in *Caenorhabditis elegans* via insulin signaling and stress response pathways. PLoS ONE 5(2):e9339

Zhang J, Lu L, Zhou L (2015) Oleanolic acid activates *daf-16* to increase lifespan in *Caenorhabditis elegans*. Biochem Biophys Res Comm 468(4):843–849

Zhang L, Jie G, Zhang J, Zhao B (2009) Significant longevity-extending effects of EGCG on *Caenorhabditis elegans* under stress. Free Rad Bio Med 46 (3):414–421

Ocimum: The Holy Basil Against Cardiac Anomalies

3

Vishnu Sharma and Debabrata Chanda

Abstract

Cardiovascular diseases have become a major health challenge in present time due to the change in lifestyle and industrialization leading to the higher morbidity and mortality. Diabetes, high blood pressure, physical inactivity, excessive weight, and consumption of alcohol and tobacco, and significant increase in life expectancy increased the frequency of cardiovascular diseases globally among all class and section of people. Cardiovascular diseases are clusters of different diseases including hypertension, coronary artery disease, angina, congestive heart disease, congenital heart disease, thrombosis, etc. Plants are being used as a remedy for various diseases since antiquity. The beneficial medicinal effects of these plant materials are mainly due to the chemical combinations of secondary metabolites produced by the plants. *Ocimum* (*Tulsi*), "Queen of herbs" has been regarded "elixir of life" and believed to promote longevity. The extract of *Ocimum* is being used in Ayurvedic remedies for common cold, headache, stomach disorder, inflammation, heart disease, malaria, etc. *Ocimum* has profound effect on treatment and prevention of cardiovascular diseases by means of lowering blood lipid content, suppressing ischemia and stroke, reducing hypertension, and also due to its higher antioxidant properties. Besides, it also plays an important role in antiplatelet aggregation and prevents risk of pulmonary hypertension. These cardioprotective properties prove that *Ocimum* may be treated as a good remedy against prevention and treatment of cardiovascular diseases.

3.1 Introduction

Cardiovascular diseases have become one of the leading causes of deaths globally (WHO 2015). There are group of diseases or injuries which alter the cardiovascular system. Generally, the people of later age life (from a range of 40–50) are more prone to cardiovascular diseases. There are many factors responsible for the development of cardiovascular diseases. Consumption of atherogenic and hypercaloric diets leads to hypertension, diabetes, dyslipidemias, overweight, and other abnormalities. Besides, diabetes is the most important risk factor to determine coronary artery disease, so that the presence of this morbidity factor is considered a risk factor equivalent to infarction, that is, despite the absence of any cardiovascular sign, diabetics are classified under

V. Sharma · D. Chanda (✉)
In Vivo Testing Facility, Molecular Bioprospection Department, Biotechnology Division, CSIR-CIMAP, Lucknow 226015, UP, India
e-mail: d.chanda@cimap.res.in

A. K. Shasany and C. Kole (eds.), *The Ocimum Genome*, Compendium of Plant Genomes, https://doi.org/10.1007/978-3-319-97430-9_3

"high cardiovascular risk" (NCEP 2002). Change in lifestyle with regard to some of these risk factors such as elevated blood sugar or blood pressure, physical inactivity, excessive weight gain, elevated lipid level, and tobacco uses or exposure to smoke may influence risk of developing cardiovascular diseases at some extent. Besides, cardiovascular diseases may arise as a result of various other factors, such as heavy physical exertion and severe emotional stress may lead to the onset of an acute situation such as acute myocardial infarction (MI), unexpected cardiac death, and stroke (Tofler and Muller 2006; Nelson et al. 2015). Chronic non-communicable diseases (CNCDs) which are comprised of cardiovascular conditions such as primarily heart disease and stroke, some types of cancers, chronic respiratory conditions, and type-2 diabetes affect individuals of all ages, nationalities, and status and are approaching an epidemic ratio worldwide. Approximately 60% of annual deaths are due to the above-stated conditions (Daar et al. 2007).

3.2 Types of Cardiovascular Diseases

Cardiovascular diseases are a cluster of different diseases which affect the cardiovascular system and lead to many physiological alterations. These include coronary heart disease, angina, congenital heart disease, peripheral atrial disease, aortic aneurysm, deep vein thrombosis, and other cardiovascular diseases.

3.2.1 Hypertension

Hypertension is a chronic medical condition in which the elevated blood pressure (systolic ($\geq$ 140 mmHg) and diastolic ($\geq$ 90 mmHg) blood pressure) of the arteries persistently increased and is also a major risk factor for coronary artery disease, stroke, heart failure, vision loss, chronic kidney disease, and peripheral vascular diseases. Recently, World Health Organization (WHO) reported that approximately 1 billion people have already got affected by hypertension globally and it is predicted to affect 1.5 billion people by the year 2025 causing more than seven million deaths annually (Guilbert 2003). In view of the severity of the condition, in every year it has been assigned for 17th day of May month as "World Hypertension Day" by the International Society of Hypertension (ISH) for focus on consideration and special attention for the awareness and management of hypertension.

3.2.2 Aortic Aneurysm/Abdominal Aortic Aneurysm

It occurs when the large blood vessel (the aorta) which supplies blood to the abdomen, pelvis, and legs becomes abnormally large outward. This is most often found in people over age 60, who have at least one or more risk factor, including emphysema, high blood pressure, high cholesterol, obesity, and smoking.

3.2.3 Acute Coronary Syndrome

This condition occurs when blood supplied to the heart is decreased or blocked, leading to a heart attack. The common symptoms of acute coronary syndrome are chest pain or discomfort, which may involve pressure, tightness, or fullness; pain or discomfort in one or both arms, jaw, neck, back, or stomach; shortness of breath; feeling dizzy or lightheaded; nausea; or sweating.

3.2.4 Angina Pectoris

Angina pectoris, also called angina, is a medical term for chest pain or discomfort due to coronary heart disease. It occurs when the heart muscle does not get as much blood as it needs. This usually happens because one or more of the heart's arteries is narrowed or blocked, also called ischemia. There are two types of angina: Stable angina refers to "predictable" chest discomfort associated with physical exertion or mental or emotional stress, while unstable angina refers to unexpected chest pain and usually occurs at rest. Unstable angina is typically more severe and prolonged.

3.2.5 Atherosclerosis

When the inner layers of artery wall become thick and irregular because of deposition of fat, cholesterol, and other substances, this accumulation of fat is then called plaque and can cause narrowing of arteries, reducing the blood flow through them. Plaque deposits can rupture, causing blood clots to form at the rupture that can block blood flow or break off and travel to another part of the body. This is a common cause of heart attack or ischemic stroke.

3.2.6 Coronary Heart Disease (CHD)/Coronary Artery Disease (CAD)

It is the most common type of heart disease. It occurs when plaque builds up in the heart's arteries, a condition called atherosclerosis. As plaque builds up, the arteries become narrow, making it more difficult for blood to flow to the heart. If blood flow to the heart becomes reduced or blocked, angina (chest pain) or a heart attack may occur. Over time, coronary artery disease can also lead to heart failure and arrhythmias.

3.2.7 Heart Attack/Acute Myocardial Infarction (AMI)

It occurs when a blocked coronary artery prevents oxygen-rich blood from reaching a section of the heart muscle. If the blocked artery is not reopened quickly, the part of the heart normally nourished by that artery begins to die. Symptoms can come on suddenly but may start slowly and persist over time. Warning signs include discomfort in the chest (pressure, squeezing, and fullness), discomfort in other upper body areas, shortness of breath, a cold sweat, nausea, or lightheadedness.

3.2.8 Congestive Heart Failure

It occurs when the heart cannot pump enough blood to the organs. The heart works, but not as well as it should. Heart failure is almost always a chronic, long-term condition. The older you are, the more common congestive heart failure becomes. Your risk also increases if you are overweight, diabetic, smoke, abuse alcohol, or use cocaine. When a heart begins to fail, fluid can pool in the body; this manifests as swelling (edema), usually in the lower legs and ankles. Fluid also may collect in the lungs, causing shortness of breath.

3.2.9 Ischemic Heart Disease (IHD)

It is a heart problem caused by heart arteries that are narrowed. When there are blockages in arteries, they become narrowed, which means less blood and oxygen reaches the heart muscle. When more oxygen is needed, such as while exercising, the heart cannot meet the demands. The lack of oxygen caused by ischemic heart disease can produce chest pain, discomfort known as angina pectoris, or even a heart attack.

3.2.10 Ischemic Stroke

It occurs when a blood clot or other particle blocks an artery in the brain or an artery leading to the brain. This causes brain cells to die or be injured. Cerebral thrombosis and cerebral embolism are ischemic strokes (American Heart Association 2013).

3.3 Plants as a Medicine

Plants are being used as potent biochemists and have been components of phytomedicine since antiquity. A heritage of knowledge on preventive and curative medicines has been found in the Atharvaveda (an Indian religious book) and Ayurveda (Indian traditional system of medicine), and so on. There are about 13,000 plant species worldwide known to have been used as drugs for the cure of different diseases. Plant-based herbal medicines can be obtained from any part of the plant like bark, leaves, flowers, roots, fruits, seeds, and so on; that is, any part of the plant may contain active components (Gordon and David 2001). The beneficial medicinal effects of plant materials consist of the chemical combinations of secondary metabolite products produced by the plant. The medicinal properties of plants are specific to particular plant species or groups as the combination of secondary products in a particular plant is taxonomically distinct (Wink 2000). The research on the medicinal plants is now being much popular worldwide. Hence, there is need of scientific examination of the remedies and standardization and quality control of the medicinal products to ensure their safety. After such safety evaluation, they can be approved for use in the primary health care. In the present chapter, the phyto-pharmaceutical aspect of one of the most important Indian medicinal plants, *Ocimum,* has been described for its medicinal uses in various cardiovascular disease conditions.

3.3.1 *Ocimum* (*Tulsi*)

Ocimum sanctum L. (also known as *Ocimum tenuiflorum*, *Tulsi*) has been used since thousands of years in Ayurveda for its diverse remedial properties. *Tulsi*, the Queen of herbs, the legendary "Incomparable one" of India, is one of the holiest and most cherished of the many healing and health-giving herbs of the orient. *Tulsi* is renowned for its religious and spiritual sanctity, as well as for its important role in the traditional Ayurvedic and Unani system of health and herbal medicine of the east (Warrier 1995). *Tulsi* is considered to be an adaptogen, balancing different processes in the body, and helpful for adapting to stress. Marked by its strong aroma and astringent taste, it is regarded in Ayurveda as a kind of "elixir of life" and believed to promote longevity. *Tulsi* extracts are used in Ayurvedic remedies for common colds, headaches, stomach disorders, inflammation, heart disease, various forms of poisoning, and malaria.

Traditionally, *O. sanctum* L. is taken in many forms, as herbal tea, extract, dried power, or fresh leaf. For centuries, the dried leaves of *Tulsi* have been mixed with stored grains to repel insects (Biswas and Biswas 2005). *O. sanctum* L. (*Tulsi*) is an erect, much-branched subshrub 30–60 cm tall, with simple opposite green or purple leaves that are strongly scented and hairy stems. Leaves have petiole and are ovate, up to 5 cm long, usually somewhat toothed. Flowers are purplish in elongate racemes in close whorls. *Tulsi* is native throughout the world tropics and widespread as a cultivated plant and an escaped weed. It is cultivated for religious and medicinal purposes and for its essential oil. *Tulsi* is an important symbol in many Hindu religious traditions, which link the plant with Goddess figure. The name "*Tulsi*" in Sanskrit means "the incomparable one." The presence of a *Tulsi* plant symbolizes the religious bend of a Hindu family.

3.3.2 Systematics of *Tulsi*

Among the plants known for medicinal value, the plants of genus *Ocimum* belonging to family Labiate are very important for their therapeutic potentials. *O. sanctum* L. (*Tulsi*), *O. gratissimum* (Ram *Tulsi*), *O. canum* (Dulal *Tulsi*), *O. bascilicum* (Ban *Tulsi*), *O. kilimandschricum, O. americanum, O. camphora,* and *O. micranthum* are examples of known important species of genus *Ocimum* that grow in different parts of the world and are known to have medicinal properties (Chopra et al. 1956; Sen 1993; Gupta et al. 2002).

3.3.3 Phytochemical Constituents

Tulsi contains vitamin C and A and minerals like calcium, zinc, and iron, as well as chlorophyll and many other phytonutrients (Anbarasu and Vijayalakshmi 2007). It also enhances the efficient digestion, absorption, and the use of nutrients from food and other herbs: Protein: 30 kcal, 4.2 g; Fat: 0.5 g; Carbohydrate 2.3 g; Calcium: 25 mg; Phosphorus 287 mg; Iron: 15.1 mg and Edible portion 25 mg vitamin C per 100 g. The chemical composition of *Tulsi* is highly complex, containing many nutrients, and other biologically active compounds, the proportions of which may vary considerably between strains and even among plants within the same field. Furthermore, the quantity of many of these constituents is significantly affected by differing growing, harvesting, processing, and storage conditions that are not yet well understood.

The nutritional and pharmacological properties of the whole herb in its natural form, as it has been traditionally used, result from synergistic interactions of many different active phytochemicals. Consequently, the overall effects of *Tulsi* cannot be fully duplicated with isolated compounds or extracts. Because of its inherent botanical and biochemical complexity, *Tulsi* standardization has, so far, eluded modern science. The leaf volatile oil contains eugenol (1-hydroxy-2-methoxy-4-allylbenzene), eugenol (also called eugenic acid), carvacrol (5-isopropyl-2-methylphenol), ursolic acid (2,3,4,5,6,6a,7, 8,8a,10,11,12,13,14b-tetradecahydro-1H-picene-4a-carboxylic acid), caryophyllene (4,11,11-trimethyl-8-methylene-bicyclo[7.2.0]undec-4-ene), linalool (3,7-dimethylocta-1,6-dien-3-ol), limatrol methyl chavicol (also called Estragole: 1-allyl-4-methoxybenzene), while the seed volatile oil have fatty acids and sitosterol; in addition, the seed mucilage contains some levels of sugars and the anthocyanins are present in green leaves (Kelm et al. 2000; Shishodia et al. 2003). The sugars are composed of xylose and polysaccharides. Although *Tulsi* is known as a general energizer and increases physical endurance, it contains no caffeine or other stimulants. The stem and leaves of holy basil contain a variety of chemical constituents that may have biological activity, including saponins, flavonoids, triterpenoids, and tannins (Jaggi et al. 2003). In addition, the following phenolic actives have been identified, which also exhibit antioxidant and anti-inflammatory activities, Rosmarinic acid ((2*R*)-2-[[(2*E*)-3-(3,4-Dihydroxyphenyl)-1-oxo-2-propenyl]]oxy]-3-(3,4-dihydroxyphenyl) propanoic acid), apigenin (5,7-dihydroxy-2-(4-hydroxyphenyl)-4H-1-benzopyran-4-one), cirsimaritin (5,4'-dihydroxy-6,7-dimethoxyflavone), isothymusin (6,7-dimethoxy-5,8,4′-trihydroxyflavone) and isothymonin. Two water-soluble flavonoids are as follows: Orientin (8- C-beta-glucopyranosyl-3′,4′, 5,7-tetrahydroxyflav-2-en-3-one) and Vicenin (6-C-beta-Dxylopyranosyl-8-C-beta-D-glucopyranosyl apigenin) (Uma Devi et al. 2000).

3.3.4 Antihypertensive Properties

Endothelial dysfunction is an important event in the development of hypertension (Cassar et al. 2003). Arterial endothelium maintains the vascular tone and regulates blood pressure and blood flow to organs and tissues (Loscalzo and Welch 1995). However, when the endothelial function is hampered, this ability of the endothelium to maintain vascular tone gets hampered. Furthermore, there is a positive correlation between hypertension and endothelial dysfunction (Rodrigo et al. 1997). Altered endothelium-mediated relaxation has been shown in patients with hypertension and in hypertensive models (Verma et al. 1996; Cardillo et al. 1998). The activation of the rennin–angiotensin system (RAS) is another most important mechanism through which hypertension leads to cardiovascular risk (Narkiewicz 2006). In the RAS, angiotensin 1-converting enzyme (ACE) (EC: 3.4.15.1) plays a major role in the regulation of blood pressure and normal cardiovascular function. It catalyzes the conversion of angiotensin I to angiotensin II, which is known to increase the blood pressure. ACE catalyzes the cleavage of angiotensin I (a decapeptide) to angiotensin II (an octapeptide) and inactivates bradykinin, a vasodilator, and hypotensive peptide (Eriksson et al. 2002). Angiotensin II stimulates vasoconstriction, increases sodium and water reabsorption, and elevates blood pressure in emergency conditions. Hence, the excessive action of ACE leads to hypertension (Lee et al. 2015); furthermore, inhibition of ACE is an important strategy for the treatment and management of hypertension (Sharma 2004; Villiger et al. 2015). Plant extracts rich in flavonoids and phenolic acids have been shown to inhibit ACE activity (Oboh et al. 2012; Irondi et al. 2016). The flavonoids are the largest group of polyphenolic compounds found in higher plants and are regarded as an excellent source of functional antihypertensive products (Croft 1998). Similarly, the phenolic acids are effective ACE inhibitors. Their ability to inhibit ACE has been attributed to the overall contribution of their functional groups, including carboxyl and hydroxyl groups; their ability to form charge–charge interactions with the zinc ion present in the active site of ACE, through the oxygen atom of their carboxylate moiety; and their interaction with the amino acids residues at the active site of ACE, to give a stable complex between the phenolic acid molecule and ACE. Thus, the ACE inhibitory activity of phenolic acids may be due to the effect of this interaction with the zinc ion and the subsequent stabilization by other interactions with amino acids in the active site (Al Shukor et al. 2013). *O. gratissimum* has stronger ACE inhibitory due to the presence of luteolin (flavonoid) and ellagic acid (phenolic acid), and luteolin had the highest ACE inhibitory activity compared to other flavonoids (Irondi et al. 2016). Similarly, *O. basilicum* extract was found to reduce systolic and diastolic blood pressure by 20 and 25 mmHg in renovascular model of hypertension in rats, which is known to have high level of angiotensin. In addition, it was also found that the extract of *O. basilicum* reduced cardiac hypertrophy and endothelin level in animal model (Umar et al. 2010). In vivo antihypertensive activity was also recorded in the essential oil of *O. gratissimum* and its major constituent eugenol at 20 mg/kg in DOCA with salt-induced hypertension in rats (Interaminense et al. 2005).

3.3.5 Effect on Ischemia

The cerebral ischemic reperfusion (IR) injury is very complex and occurs through a series of mechanisms of which oxidative stress is considered a major culprit for neurodegeneration (Warner et al. 2004). Role of oxidative stress in the development of IR-mediated cerebral injury is well documented (Nour et al. 2013). Among all regions of the brain, hippocampus is highly vulnerable to deleterious effects of ROS. Excessive ROS causes damage to biomolecules like DNA, lipids, and proteins (Chen et al. 2011). Fall in endogenous antioxidants (GSH and SOD activity) is well documented in cerebral IR injury (Surapaneni et al. 2016). Increase in thiobarbituric acid reactive substances (TBARS) levels, a measure of malondialdehyde produced by lipid membrane damage due to free radicals is an indication of cerebral IR (Margaill et al. 2005). *O. basilicum* extract is rich in phenolic compounds like p-hydroxy benzoic acid (0.14%), caffeic acid (0.28%), chlorogenic acid (0.13%), cinnamic acid (0.09%), p-coumaric acid (0.18%), ferulic acid (0.08%), and gallic acid (0.03%) along with traces of vanillic acid. Phenolic acids have a profound antioxidant and thus have a potent neuroprotective role. Administration of ferulic acid was shown to inhibit cerebral infarction, free radical-induced apoptosis, inflammation, and oxidative stress in animal model of cerebral ischemia (Cheng et al. 2008). Thus, the levels of TBARS, reduced GSH, and SOD activity were measured to estimate the extent of ROS damage in IR-mediated injury. The restorative effects of *O. basilicum* and *O. gratissimum* extracts on brain endogenous antioxidants levels against IR-induced cerebral injury have been investigated and revealed that phenolic compounds including flavonoids are well documented for antioxidant as well as neuroprotective properties (Kelsey et al. 2010; Bora et al. 2011). The antioxidant and neuroprotective role of rosmarinic acid, caffeic acid, ferulic acid, vanilic acid, quercetin, and rutin shows a linear relationship between total phenol content and antioxidant activity of *O. kilimandscharicum* and may contribute its efficacy in ischemia (Hakkim et al. 2008).

3.3.6 Effect on Pulmonary Hypertension

Pulmonary hypertension (PH) is defined and diagnosed hemodynamically as mean pulmonary artery pressure (mPAP) of more than 25 mm of Hg at rest along with normal pulmonary capillary wedge pressure and pulmonary vascular resistance of more than 3 woods unit after right heart catheterization (Simonneau et al. 2013). *Ocimum* has therapeutic potential against structural changes in pulmonary arteries. This effect of *Ocimum* affirms its disease-modifying effect and has a significant clinical relevance in pulmonary vascular remodeling, a pathological feature of PH which leads to increased pulmonary vascular resistance (Jeffery and Wanstall 2001; Meghwani et al. 2017), rise in pulmonary artery pressure and vascular hypertrophy (Bogdan et al. 2012). The oxidative stress plays an important role in the development and progression of PH. Nicotinamide adenine dinucleotide phosphate (NADPH) oxidase (NOX) has been demonstrated to be the major source for reactive oxygen species (ROS). In addition, NOX-1 expression has been found to be increased in human and rat PH, respectively (Csiszar et al. 2009; Ghouleh et al. 2017). Moreover, it has also been reported that NOX-derived ROS are vital modulators of signal transduction pathways that control key physiological activities such as cell growth, proliferation, and apoptosis (Manea et al. 2015). Therefore, the reduced level of NOX could be beneficial for morbidity and mortality in the patients with PH (McLaughlin et al. 2011; Korsholm et al. 2015). In addition, *Ocimum* had an anti-apoptotic effect and preserved the decrease in Bcl2/Bax ratio in renovascular hypertrophy secondary to the development of PH. *Ocimum* has apparent anti-apoptotic effect which is useful in preventing renovascular failure, a major cause of morbidity in PH (Meghwani et al. 2018).

3.3.7 Lipid-Lowering Properties of *Ocimum*

It has been widely known that enhancement of serum cholesterol level can lead to atherosclerosis; blood supply to the organs gradually compromises until organ function becomes impaired. The aqueous leaf extract of *O. sanctum* decreased the hepatic lipid content and the elevation of fecal bile acid excretion without the change of fecal lipid excretion by means of increasing hepatic bile acids biosynthesis using cholesterol as the precursor, and finally leading to the decrease of hepatic cholesterol and triglyceride accumulations which indicate the hepatic lipid-lowering effect of *Ocimum,* and it is probably related to a lower intestinal bile acids absorption (Suanarunsawat et al. 2011). *Ocimum* was found not only to lower the serum and hepatic lipid but also to suppress the high serum levels of AST, ALT, LDH, and CK-MB. It is also found to suppress lipid peroxidation and increases the activities of antioxidant enzymes in both the liver and cardiac tissues. *Ocimum* had a free radical-scavenging activity which probably provides organs protection from hypercholesterolemia and several lines of evidence showed that plants with phenolic compounds had antilipidemic and antioxidative activities to protect the liver and heart against hyperlipidemia (Auger et al. 2002; Yokozawa et al. 2006). Higher levels of TGs and cholesterol are one of the risk factors of CAD in humans (Gupta et al. 2006; Talayero et al. 2011). Thus, reduction in cholesterol and TGs seen in human population is a very positive outcome for reduction of dyslipidemia-related CVD risks (Saggini et al. 2011). This reinforces the traditional claim that *Tulsi* is good for heart.

3.3.8 Antiplatelet Aggregation Activity

Endothelial dysfunction and platelet aggregation play important role in the development of cardiovascular diseases and the major complications of atherosclerosis and arterial thrombus formation (Grover-Paez and Zavalza-Gomez 2009). Platelet aggregation is critical for the progression of thromboembolic diseases pathogenesis (Cassar et al. 2003). Hence, by targeting platelet aggregation with drugs and/or dietary intervention can prevent or treat thrombosis and reduce the related cardiovascular disease (La Rosa et al. 1990). The polyphenol-rich diets have been shown to play a positive role in vascular functioning including platelet aggregation in humans (Murphy et al. 2003). In fact, recent studies on platelet function demonstrated that the phenolic compounds isolated from medicinal plants are responsible for their beneficial effect and attenuate the platelet aggregation (Mekhfi et al. 2006). The aqueous basil extract inhibits platelet aggregation and thrombin-induced platelet activation. *O. basilicum* exerts in vivo antiplatelet and antithrombotic effects (Tohti et al. 2006). Also, the dose dependency of *O. basilicum* effect pleads in favor of the reality of the activity (Amrania et al. 2009). *O. sanctum* fixed oil also increases blood clotting time, and the response was comparable to that obtained with aspirin. The effect appears to be due to the antiaggregatory action of oil on platelets. Linolenic acid contained in the oil can be metabolised to EPA, which can inhibit the formation of TXA2 through cyclooxygenase and produce PGI3 and TXA3. Like PGI2, PGI3 also possesses antiaggregatory property while TXA3 has much less proaggregatory activity toward platelets compared with TXA2 (Zurier 1991; Fischer and Weber 1984; Lee et al. 1985). Combined antiaggregatory effects of PGI2 and PGI3 supplemented by inhibition of TXA2 could, therefore, contribute toward anticoagulant effect of *O. sanctum* fixed oil. *O. sanctum* fixed oil contains EFAs like linoleic and linolenic acids. Linoleic acid is a precursor of gamma-linolenic acid (GLA) and arachidonic acid which can competitively inhibit formation of cyclooxygenase products (PGE2, TXA2) from A.A. EPA produces TXA3 (which has much lesser ability than TXA2 to constrict blood vessels) and PGI3 which has vasodilatory activity (Fischer and Weber 1984; Lee et al. 1985; Zurier 1991).

3.3.9 Antioxidant and Antidiabetic Properties

Free radical-scavenging enzymes such as catalase, superoxide dismutase, and glutathione peroxidase are the first-line cellular defense against oxidative injury, decomposing O_2 and H_2O_2 before their interaction to form the more reactive hydroxyl radical (OH^*). The equilibrium between these enzymes is an important process for the effective removal of oxygen stress in intracellular organelles. The second line of defense consists of the non-enzymatic scavenger's viz. ascorbic acid, α-Tocopherol, ceruloplasmin, and sulphydryl containing compounds, which scavenge residual free radicals escaping decomposition by the antioxidant enzymes. Besides, antioxidant enzymes and physiological antioxidants, alteration in LDH has been considered as one of the most important markers of myocardial infarction. Glutathione is implicated in the removal of free oxygen species such as H_2O_2, superoxide radicals, alkoxy radicals, and maintenance of membrane protein thiols and as a substrate for glutathione peroxidase (GPX) and glutathione-S transferase (GST). Superoxide anion either can be generated by reduction of one electron from molecular oxygen or oxidation of one electron of hydrogen peroxide. A significant number of enzymatic reactions in biological systems result in the creation of this radical, and the highest amounts are produced in reactions of oxidases such as xanthine oxidase (XOD) and aldehyde oxidase, but also reactions catalyzed by NADPH-cytochrome C reductase, NADPH-cytochrome P_{450} reductase, etc. In the reaction with H_2O_2, superoxide anion radical produces hydroxyl ion (Haber-Weiss or Fenton reaction), while the reaction with nitricoxide formed peroxynitrite anion (ONOO–) which may have greater toxicity than the extracellular OH radical (Đorđević et al. 2000). *Ocimum* enhances the levels of GSH by virtue of its increased synthesis or due to improved glutathione reductase activity. *Ocimum* inhibits the LP, enhancement of SOD activity and improvement in GSH levels and leads to the cardioprotection by means of its antioxidant properties. Antioxidant properties of *Ocimum* are due to its constituents like eugenol, flavonols, flavones, and anthocyanins. The *Ocimum*-induced decrease in SOD levels may be due to the involvement of superoxide free radical in myocardial cell damage. A decrease in activity of SOD can result in the decreased removal of superoxide ion, which can be harmful to the myocardium (Liu et al. 1977). It is possible that in the presence of *Ocimum* either generation of free radical itself is impaired or enhanced; SOD activity could effectively scavenge the first free radical superoxide from the system. The *Ocimum* significantly protects the heart by maintaining the LDH levels is an indicative of the fact that *Ocimum* has cardioprotective action and maintain membrane integrity of myocytes. *O. basilicum* extracts showed a strong effect on the inhibition of superoxide anion radical, even at very low applied concentrations. The activity of the extracts obtained from *O. basilicum* was approximately equal to that of the synthetic antioxidants. Hydroxyl radical (OH^-) is chemically the most reactive form of "activated oxygen," which occurs during the reduction of molecular oxygen and is responsible for the cytotoxic effects of oxygen (Imlay and Linn 1988). Due to its extreme reactivity, hydroxyl radicals react immediately with biomolecules and can effectively "attack" every molecule present in living cells such as sugars, amino acids, phospholipids, pyrimidine and purine bases, and organic acids. The OH radical scavenging capacity of *O. basilicum* extract exhibited different behavior with respect to the production of OH radicals. Since lipid peroxidation causes oxidative damage to cell membranes and all other systems that contain lipids, in any investigation of total antioxidative activity of extracts it is necessary to investigate their effects on lipid peroxidation (Chatterjee and Agarwal 1988). The high inhibitory effects of *Ocimum* extract can be related to the presence of the amount of total phenolic contents and content of total flavonoids in the extracts. It has already been reported that flavonoids act as powerful scavengers of free radicals (Robak and Gryglewski 1988). Different flavonoids inhibit lipid peroxidation in vitro, and the most pronounced effect is exhibited by quercetin (Husain et al. 1987).

3.4 Summary

Cardiovascular disease is a major cause of morbidity and premature death throughout the world and a challenge for health care. The underlying pathology is very broad and time-dependent. While atherosclerosis takes several years to develop, acute coronary and cerebrovascular events frequently occur suddenly and are often fatal before medical care can be given. Modification of some risk factors has been shown to reduce mortality and morbidity in people suffering from cardiovascular disease. Several forms of therapy have now been evolved to prevent coronary, cerebral, and peripheral vascular events. Phytomedicines are one of them, they provide good health without any adverse effects as their therapeutic index is much higher than conventional medicines. *O. sanctum* Linn. (*Tulsi*) is a well-known plant that is grown all over India. Several medicinal properties have been attributed to the plant in Indian traditional medicine, in which it is mostly used as an aqueous extract of leaves. The leaf juice has been used for chronic fever, hemorrhage, dysentery, dyspepsia, and skin diseases. It contains good antioxidant activity, and this has been mainly attributed to the presence of compounds such as flavonoids, tannins, ascorbic acid, and carotenoids. Its oil and leaf extract has significant anti-inflammatory, antipyretic, analgesic, antiarthritic, anticoagulant, hypotensive, antibacterial, and chemoprotective activities. It has been reported that *Ocimum* has profound efficacy against the cardiovascular diseases and has significant effect on various cardiovascular diseases including lipid-lowering effect which leads to a low risk of developing atherosclerosis, antihypertensive effects on blood pressure by means of inhibiting angiotensin II, anti-apoptotic activity of *Ocimum* prevents the pulmonary hypertension, besides, inhibition of NOX-1 gene leads to prevention of pulmonary hypertension. Further, *Ocimum* inhibits the formation of cyclooxygenase and stimulates the biosynthesis of TXA3 and PGI3 which has vasodilatory activity. The leaf extract of *Ocimum* has also possessed antioxidative properties which reduces the risk of Ischemia and other cardiovascular diseases. Hence, *Ocimum* can be used as herbal medicine for the treatment of cardiovascular diseases but further there is a need for series of studies in the fulfillment of this comment.

References

Al Shukor N, Van Camp J, Gonzales GB, Staljanssens D, Struijs K, Zotti MJ (2013) Angiotensin-converting enzyme inhibitory effects by plant phenolic compounds: a study of structure activity relationships. J Agric Food Chem 61:11832–11839

American Heart Association (2013) About Cardiac Arrest, http://www.heart.org/HEARTORG/Conditions/More/CardiacArrest/About-Cardiac-Arrest_UCM_307905_Article.jsp

Amrania S, Harnafib H, Gadib D, Mekhfib M, Legssyerb A, Azizb M, Martin-Nizardc FO, Bosca L (2009) Vasorelaxant and anti-platelet aggregation effects of aqueous *Ocimum basilicum* extract. J Ethnopharmacol 125:157–162

Anbarasu K, Vijayalakshmi G (2007) Improved shelf life of protein-rich tofu using *Ocimum sanctum* (*Tulsi*). extracts to benefit Indian rural population. J Food Sci 72:M300–M305

Auger C, Caporiccio B, Landrault N, Teissedre PL, Laurent C, Cros G, Besançon P, Rouanet JM (2002) Red wine phenolic compounds reduce plasma lipids and apolipoprotein B and prevent early aortic atherosclerosis in hypercholesterolemic golden Syrian hamsters (*Mesocricetus auratus*). J Nutr 132(6): 1207–1213

Biswas NP, Biswas AK (2005) Evaluation of some leaf dusts as grain protectant against rice weevil *Sitophilus oryzae* (Linn.). Environ Ecol 23:485–488

Bogdan S, Seferian A, Totoescu A, Dumitrache-Rujinski S, Ceausu M, Coman C, Ardelean CM, Dorobantu M, Bogdan M (2012) Sildenafil reduces inflammation and prevents pulmonary arterial remodeling of the monocrotaline-induced disease in the wistar rats. Maedica 7:109–116

Bora KS, Shri R, Monga J (2011) Cerebroprotective effect of *Ocimum gratissimum* against focal ischemia and reperfusion-induced cerebral injury. Pharm Biol 49:175–181

Cardillo C, Kilcoyne CM, Quyyumi AA, Cannon RO III, Panza JA (1998) Selective defect in nitric oxide synthesismay explain the impaired endotheliumdependent vasodilatation in patients with essential hypertension. Circulation 97:851–856

Cassar K, Bachoo P, Brittenden J (2003) The role of platelets in peripheral vascular disease. Eur J Vasc Endovasc Surg 25:6–15

Chatterjee N, Agarwal S (1988) Liposomes as membrane model for study of lipid peroxidation. Free Radic Biol Med 4:51–72

Chen H, Yoshioka H, Kim GS, Jung JE, Okami N, Sakata H, Maier CM, Narasimhan P, Goeders CE, Chan PH (2011) Oxidative stress in ischemic brain damage: mechanisms of cell death and potential molecular targets for neuroprotection. Antioxid Redox Signal 14:1505–1517

Cheng CY, Su SY, Tang NY, Ho TY, Chiang SY, Hsieh CL (2008) Ferulic acid provides neuroprotection against oxidative stress-related apoptosis after cerebral ischemia/reperfusion injury by inhibiting ICAM-1 mRNA expression in rats. Brain Res 13:136–150

Chopra RN, Nayer SI, Chopra IC (1956) Glossary of Indian medicinal plants. CSIR, New Delhi

Croft KD (1998) The chemistry and biological effects of flavonoids and phenolic acids. Ann NY Acad Sci 854:435–442

Csiszar A, Labinskyy N, Olson S, Pinto JT, Gupte S, Wu JM, Hu F, Ballabh P, Podlutsky A, Losonczy G (2009) Resveratrol prevents monocrotaline-induced pulmonary hypertension in rats. Hypertension 54:668–675

Daar AS, Singer PA, Persad DL, Pramming SK, Matthews DR, Beaglehole R, Bernstein A, Borysiewicz LK, Colagiuri S, Ganguly N, Glass RI (2007) Grand challenges in chronic non-communicable diseases. Nature 450(7169):494–496

Đorđević VB, Pavlović DD, Kocić GM (2000) Biohemija Slobodnih Radikala. Niš, Serbia, Medicinski Fakultet, pp 132–138

Eriksson U, Danilczyk U, Penninger JM (2002) Just the beginning: novel functions for angiotensin-converting enzymes. Curr Biol 12:R745–R752

Fischer S, Weber PC (1984) Prostaglandin I3 is formed in vivo in man after dietary eicosapentaenoic acid. Nature 307:165–184

Ghouleh IA, Sahoo S, Meijles DN, Amaral JH, de Jesus DS, Sembrat J, Rojas M, Goncharov DA, Goncharova EA, Pagano PJ (2017) Endothelial Nox1 oxidase assembly inhuman pulmonary arterial hypertension; driver of gremlin1-mediated proliferation. Clin Sci 131:2019–2035

Gordon MC, David JN (2001) Naturan product drug discovery in the next millennium. Pharm Boil 39:8–17

Grover-Paez F, Zavalza-Gomez AB (2009) Endothelial dysfunction and cardiovascular risk factors. Diabetes Res Clin Pract 84:1–10

Guilbert JJ (2003) The World health report 2002 - reducing risks, promoting healthy life. Educ Health 16 (2): 230 (Abingdon, England)

Gupta SK, Prakash J, Srivastav S (2002) Validation of traditional claim of *Tulsi*, *Ocimum sanctum* Linn. as a medicinal plant. Indian J Exp Biol 40:765–773

Gupta S, Mediratta PK, Singh S, Sharma KK, Shukla R (2006) Antidiabetic, antihypocholestrolaemic and antioxidant effect of *Ocimum sanctum* (Linn) seed oil. Indian J Exp Biol 44:300–304

Hakkim FL, Arivazhagan G, Boopathy R (2008) Antioxidant property of selected *Ocimum* species and their secondary metabolite content. J Med Plants Res 2:250–257

Husain R, Cilliard J, Cilliard P (1987) Hydroxyl radical scavenging activity of flavonoids. Phytochemistry 26:2489–2491

Imlay JA, Linn S (1988) DNA damage and oxygen radical toxicity. Science 240:1302–1309

Interaminense LF, Leal-Cardoso JH, Magalhães PJ, Duarte GP, Lahlou S (2005) Enhanced hypotensive effects of the essential oil of *Ocimum gratissimum* leaves and its main constituent, eugenol, DOCA-salt hypertensive conscious rats. Planta Med 71(4):376–378

Irondi EA, Agboola SO, Oboh G, Boligon AA (2016) Inhibitory effect of leaves extracts of *Ocimum basilicum* and *Ocimum gratissimum* on two key enzymes involved in obesity and hypertension *in vitro*. J Intercult Ethnopharmacol 5(4):396–402

Jaggi RK, Madaan R, Singh B (2003) Anticonvulsant potential of holy basil, *Ocimum sanctum* Linn., and its cultures. Indian J Exp Biol 41:1329–1333

Jeffery TK, Wanstall JC (2001) Pulmonary vascular remodeling: a target for therapeutic intervention in pulmonary hypertension. Pharmacol Ther 92:1–20

Kelm MA, Nair MG, Strasburg GM, DeWitt DL (2000) Antioxidant and cyclooxygenase inhibitory phenolic compounds from *Ocimum sanctum* Linn. Phytomedicine 7:7–13

Kelsey NA, Wilkins HM, Linseman DA (2010) Nutraceutical antioxidants as novel neuroprotective agents. Molecules 15:7792–7814

Korsholm K, Andersen A, Kirkfeldt RE, Hansen KN, Mellemkjær S, Nielsen-Kudsk JE (2015) Survival in an incident cohort of patients with pulmonary arterial hypertension in Denmark. Pulm Circ 5:364–369

La Rosa JC, Hunninghake D, Bush D (1990) The cholesterol fact: a summary of the evidence relating dietary fats, serum cholesterol and CHD. A joint statement by the American Heart Association and the National Heart, Lung and Blood Institute. Circulation 81:1721–1733

Lee TH, Hoover RL, Williams JD, Sperling RI, Ravalese JR III, Spur BW, Robinson DR, Corey EJ, Lewis RA, Austen KF (1985) Effect of dietary enrichment with eicosapentaenoic and docosahexaenoic acids on in vitro neutrophil and monocyte leucotriene generation and neutrophil function. New England J Med 312:1217–1224

Lee BH, Lai YS, Wu SC (2015) Antioxidation, angiotensin converting enzyme inhibition activity, nattokinase, and antihypertension of *Bacillus subtilis* (natto)-fermented pigeon pea. J Food Drug Anal 1:1–8

Liu J, Simon LM, Philips JR, Robin ED (1977) Superoxide dismutase (SOD) activity in hypoxic mammalian systems. J Appl Physiol 42:107–110

Loscalzo J, Welch G (1995) Nitric oxide and its role in the cardiovascular system. Prog Cardiovasc Dis 38:87–104

Manea SA, Constantin A, Manda G, Sasson S, Manea A (2015) Regulation of Nox enzymes expression in vascular pathophysiology: focusing on transcription factors and epigenetic mechanisms. Redox Biol 5:358–366

Margaill I, Plotkine M, Lerouet D (2005) Antioxidant strategies in the treatment of stroke. Free Radic Biol Med 39:429–443

McLaughlin VV, Davis M, Cornwell W (2011) Pulmonary arterial hypertension. Curr Probl Cardiol 36:461–517

Meghwani H, Prabhakar P, Mohammed SA, Seth S, Hote MP, Banerjee SK, Arava S, Ray R, Maulik SK (2017) Beneficial effects of aqueous extract of stem bark of *Terminalia arjuna* (Roxb.), An ayurvedic drug in experimental pulmonary hypertension. J Ethnopharmacol 197:184–194

Meghwani H, Prabhakar P, Mohammed SA, Dua P, Seth S, Hote MP, Banerjee SK, Arava S, Ray R, Maulik SK (2018) Beneficial effect of *Ocimum sanctum* (Linn) against monocrotaline-induced pulmonary hypertension in rats. Medicines 5(34):1–15

Mekhfi H, ElHaouari M, Bnouham M, Aziz M, Ziyyat A, Legssyer A (2006) Effects of extracts and tannins from *Arbutus unedo* leaves on rat platelet aggregation. Phytother Res 20:135–139

Murphy KJ, Chronopoulos AK, Singh I, Francis MA, Moriarty H, Pike MJ, Turner AH, Mann NJ, Sinclair AJ (2003) Dietary flavanols and procyanidin oligomers from cocoa (*Theobroma cacao*) inhibit platelet function. Am J Clin Nutr 77:1466–1473

Narkiewicz K (2006) Obesity and hypertension - the issue is more complex than we thought. Nephrol Dial Transplant 21:264–267

National Cholesterol Education Program (NCEP III) (2002) Third report of the National Cholesterol Education Program (NCEP) expert panel on detection, evaluation and treatment of high blood cholesterol in adults (adult treatment panel III) final report 2002. Circulation 106(25):3143–3421

Nelson F, Nyarko KM, Binka FN (2015) Prevalence of risk factors for non-communicable diseases for new patients reporting to Korle-Bu Teaching Hospital. Ghana Med J 49(1):12–18

Nour M, Scalzo F, Liebeskind DS (2013) Ischemia-reperfusion injury in stroke. Interv Neurol 1:185–199

Oboh G, Ademiluyi AO, Akinyemi AJ, Henle T, Saliu JA, Schwarzenbolz U (2012) Inhibitory effect of polyphenol-rich extracts of Jute leaf (*Corchorus olitorius*) on key enzyme linked to type-2 diabetes (α-amylase and α-glucosidase) and hypertension (angiotensin I converting) in vitro. J Funct Food 4:450–458

Robak J, Gryglewski J (1988) Flavonoids are scavengers of superoxide anion. Biochem Pharmacol 37:837–841

Rodrigo E, Maeso R, Muanoz-Garcaia R (1997) Endothelial dysfunction in spontaneously hypertensive rats: consequences of chronic treatment with losartan or captopril. J Hypertens 15:613–618

Saggini A, Anogeianaki A, Angelucci D, Cianchetti E, D'Alessandro M, Maccauro G, Salini V, Caraffa A, Teté S, Conti F, Tripodi D, Fulcheri M, Frydas S, Shaik-Dasthagirisaheb YB (2011) Cholesterol and vitamins: revisited study. J Biol Regul Homeost Agents 25(4):505–515

Sen P (1993) Therapeutic potential of *Tulsi*: from experience to facts. Drug News Views 1:15–21

Sharma AM (2004) Is there a rationale for angiotensin blockade in the management of obesity hypertension? Hypertension 44:12–19

Shishodia S, Majumdar S, Banerjee S, Aggarwal BB (2003) Urosolic acid inhibits nuclear factor-kappa B activation induced by carcinogenic agents through suppression of I kappa B alpha kinase and p65 phosphorylation: correlation with down-regulation of cyclooxygenase 2, matrix metalloproteinase 9, and cyclin D1. Cancer Res 63:4375–4383

Simonneau G, Gatzoulis MA, Adatia I, Celermajer D, Denton C, Ghofrani A, Gomez Sanchez MA, Krishna Kumar R, Landzberg M, Machado RF (2013) Updated clinical classification of pulmonary hypertension. J Am Coll Cardiol 62:D34–D41

Suanarunsawat T, Ayutthaya WDN, Songsak T, Thirawarapan S, Poungshompoo S (2011) Lipid-lowering and antioxidative activities of aqueous extracts of *Ocimum sanctum* L. leaves in rats fed with a high-cholesterol diet. Oxid Med Cell Longev. https://doi.org/10.1155/2011/962025

Surapaneni S, Prakash T, Ansari M, Manjunath P, Kotresha D, Goli D (2016) Study on cerebroprotective actions of *Clerodendron glandulosum* leaves extract against long term bilateral common carotid artery occlusion in rats. Biomed Pharmacother 80:87–94

Talayero BG, Sacks FM (2011) The role of triglycerides in atherosclerosis. Curr Cardiol Rep 13(6):544–552

Tofler GH, Muller JE (2006) Triggering of acute cardiovascular disease and potential preventive strategies. Circulation 114(17):1863–1872

Tohti I, Tursun M, Umar A, Subat T, Imin H, Moore N (2006) Aqueous extracts of *Ocimum basilicum* L (sweet basil) decrease platelet aggregation induced by ADP and thrombin in vitro and rats arterio-venous shunt thrombosis in vivo. Thromb Research 118:733–739

Uma Devi P, Ganasoundari A, Vrinda B, Srinivasan KK, Unnikrishnan MK (2000) Radiation protection by the *Ocimum* flavonoids orientin and vicenin: Mechanisms of action. Radiat Res 154:455–460

Umar A, Imam G, Yimin W, Kerim P, Tohti I, Berké B, Moore N (2010) Antihypertensive effects of *Ocimum basilicum* L. (OBL) on blood pressure in renovascular hypertensive rats. Hypertens Res 33(7):727–730

Verma S, Bhanot S, Yao L, McNeill J (1996) Defective endothelium-dependent relaxation in fructose-hypertensive rats. Am J Hypertens 9:370–376

Villiger A, Sala F, Suter A, Butterweck V (2015) In vitro inhibitory potential of *Cynara scolymus*, *Silybum marianum*, *Taraxacum officinale*, and *Peumus boldus* on key enzymes relevant to metabolic syndrome. Phytomedicine 22:138–144

Warner DS, Sheng H, Batinić-Haberle I (2004) Oxidants, antioxidants and the ischemic brain. J Exp Biol 207:3221–3231

Warrier PK (1995) In: Longman O (ed) Indian medicinal plants. CBS publication, New Delhi, p 168

Wink M (2000) Introduction biochemistry, role and biotechnology of secondary products. In: Wink M (ed) Biochemistry of secondary product metabolism. CRC press, Boca Raton, Florida, pp 1–16

World Health Organization (2015) Cardiovascular diseases (CVDs), http://www.who.int/mediacentre/factsheets/fs317/en/

Yokozawa T, Cho EJ, Sasaki S, Satoh A, Okamoto T, Sei Y (2006) The protective role of Chinese prescription Kangenkaryu extract on diet-induced hypercholesterolemia in rats. Biol Pharmaceut Bull 29 (4):760–765

Zurier B (1991) Essential fatty acids and inflammation. Ann Rheumat Dis 50(11):745–746

4 Evolutionary Analysis of a Few Protein Superfamilies in *Ocimum tenuiflorum*

A. Gandhimathi, Nitish Sathyanarayanan, Meenakshi Iyer, Rachit Gupta and R. Sowdhamini

Abstract

Phytochemicals in the form of secondary metabolites produced by plants have been used for therapeutic purposes, some of the well-known examples being artemisinin for treatment of malaria, vinblastine and vinblastine and vincristine for treatment of cancer. Plants produce several such secondary metabolites having anticancer, cardioprotectant, anti-inflammatory, antidiabetic, artificial sweetener, antimicrobial properties, and plants have evolved elaborate pathways to synthesize these complex biomolecules. Some of these molecules can be highly complex in their chemistry, and it is often impossible to synthesize them in the laboratory, while plants have evolved enzymes with a remarkable capacity to catalyze these reactions with chemo-, regio-, and stereospecificity. Understanding sequence and structural properties of plant enzymes involved in the synthesis of metabolites will help in deciphering the mechanism underlying the synthesis of these phytochemicals. In the present chapter, we describe a computational pipeline for identifying, validating, and analyzing the key components involved in the synthesis of terpenoids and a less studied class of proteases called rhomboids. A bioinformatic study of this nature will have wider implication as not only a tool to understand sequence and structure–function relationships of some of the well-studied metabolites and enzymes, to aid protein engineering for biotechnological utilization of these commercially valuable molecules.

4.1 Introduction

Plant enzymes are critical for the production of phytochemicals such as secondary metabolites. Although more than 25,000 plant species are cataloged as medicinally important (Farnsworth 1988), the understanding of the biochemical basis for these properties is extremely limited. Moreover, of the million metabolites estimated to be synthesized by plants (Afendi et al. 2012), the biosynthetic pathways are known for only about 0.1% (Caspi et al. 2016). Advancements in sequencing techniques and development of resources for big data analysis have resulted in large numbers of plant genome sequencing projects being undertaken and an exponential growth of biological sequences. Annotating a

A. Gandhimathi · N. Sathyanarayanan · M. Iyer
R. Gupta · R. Sowdhamini (✉)
National Centre for Biological Sciences, GKVK Campus, Bellary Road, Bangalore 560065, India
e-mail: mini@ncbs.res.in

R. Gupta
Institute of Chemical Technology, Mumbai, India

A. K. Shasany and C. Kole (eds.), *The Ocimum Genome*, Compendium of Plant Genomes, https://doi.org/10.1007/978-3-319-97430-9_4

gene product to characterize its structure and function in the laboratory is a lengthy and expensive process. Increasingly, in silico methods are being used to annotate the genes that encode for these plant biosynthetic pathways (Radivojac et al. 2013). Understanding how enzyme superfamilies evolve is vital for accurate genome annotation, predicting protein functions, and protein engineering. The sequence and functional diversity of enzyme superfamilies have expanded through billions of years of evolution from a common ancestor. Integrative approaches that examine protein sequence, structure, and function have begun to provide comprehensive views of the functional diversity and evolutionary relationships within enzyme superfamilies. In the present chapter, we explore the sequence and structural space for biosynthesis of terpenoids and cytochrome P450s, which form one of the largest classes of plant metabolites. We have also explored rhomboids, which are ubiquitous intramembrane serine proteases, found in nearly all sequenced genomes, including plants. Although the class of enzymes is found to be abundant in plants, their roles in plants remain elusive.

4.1.1 Biosynthesis of Terpenoids

Terpenes (also known as terpenoids or isoprenoids) are one of the largest group of plant metabolites, with more than 30,000 different structures (Kong et al. 2011), spread over the widest assortment of structural types with hundreds of different monoterpene, sesquiterpene, diterpene, and triterpene carbon skeletons (Degenhardt et al. 2009). The majority of terpenes have been isolated from plants, where their enormous structural variability leads to a great functional diversity. Terpenes play important roles in almost all basic plant processes, including growth, development, reproduction and defense. Many terpenoids have reported therapeutic properties such as antimicrobial, anti-inflammatory, immunomodulatory, and chemotherapeutic properties, making them highly interesting in the medical field. Also, they are widely used in the flavors and fragrances in industries, in addition to being a source of biofuels (Pazouki and Niinemets 2016). Terpenoids are derived by repetitive fusion of the five-carbon units based on isopentane skeleton (Fig. 4.1). The terpenoidal backbone is synthesized from the two precursors: isopentenyl pyrophosphate (IPP) and dimethylallyl pyrophosphate (DMAPP) through a different number of repeats, rearrangement, and cyclization reactions (Abdallah and Quax 2017). On the basis of C_5 units, we can classify the terpenoids as C_5 (hemiterpenes), C_{10} (monoterpenes), C_{15} (sesquiterpenes), C_{20} (diterpenes), C_{25} (sesterpenes), C_{30} (triterpenes), C_{40} (tetraterpenes), and $>C_{40}$ (polyterpenes) (Martin et al. 2003).

4.1.2 Terpene Synthases

Terpene synthases/cyclases (TPS) are a family of enzymes responsible for synthesizing the vast array of terpenoid compounds. These enzymes are involved in catalyzing the rearrangement and/or cyclization of the precursors geranyl diphosphate (GPP), farnesyl diphosphate (FPP), and geranylgeranyl diphosphate (GGPP) to produce the different classes of terpenoids (Tholl 2006). TPS genes are generally divided into seven clades, with some plant lineages having a majority of their TPS genes in one or two clades. This indicates lineage-specific expansion of specific types of genes (Table 4.1). The TS genes were classified into the terpene synthase (TPS) family, and into more specific subfamilies, or as triterpene cyclases (TTCs), or squalene synthases (SSs) (Boutanaev et al. 2015). The interesting structural diversity of terpenoids is based on the orientation of the substrate in the active site, which undergoes a series of cyclizations and/or rearrangements to produce a certain terpenoid by the associated TPS (Bohlmann et al. 1998). TPS are generally classified into class I and class II TPS based on their substrate activation mechanisms. Class I TPS are characterized by catalyzing the ionization of the allylic diphosphate ester bond in their isoprenyl substrates, while class II TPS catalyze protonation-induced cyclization reaction of the substrate

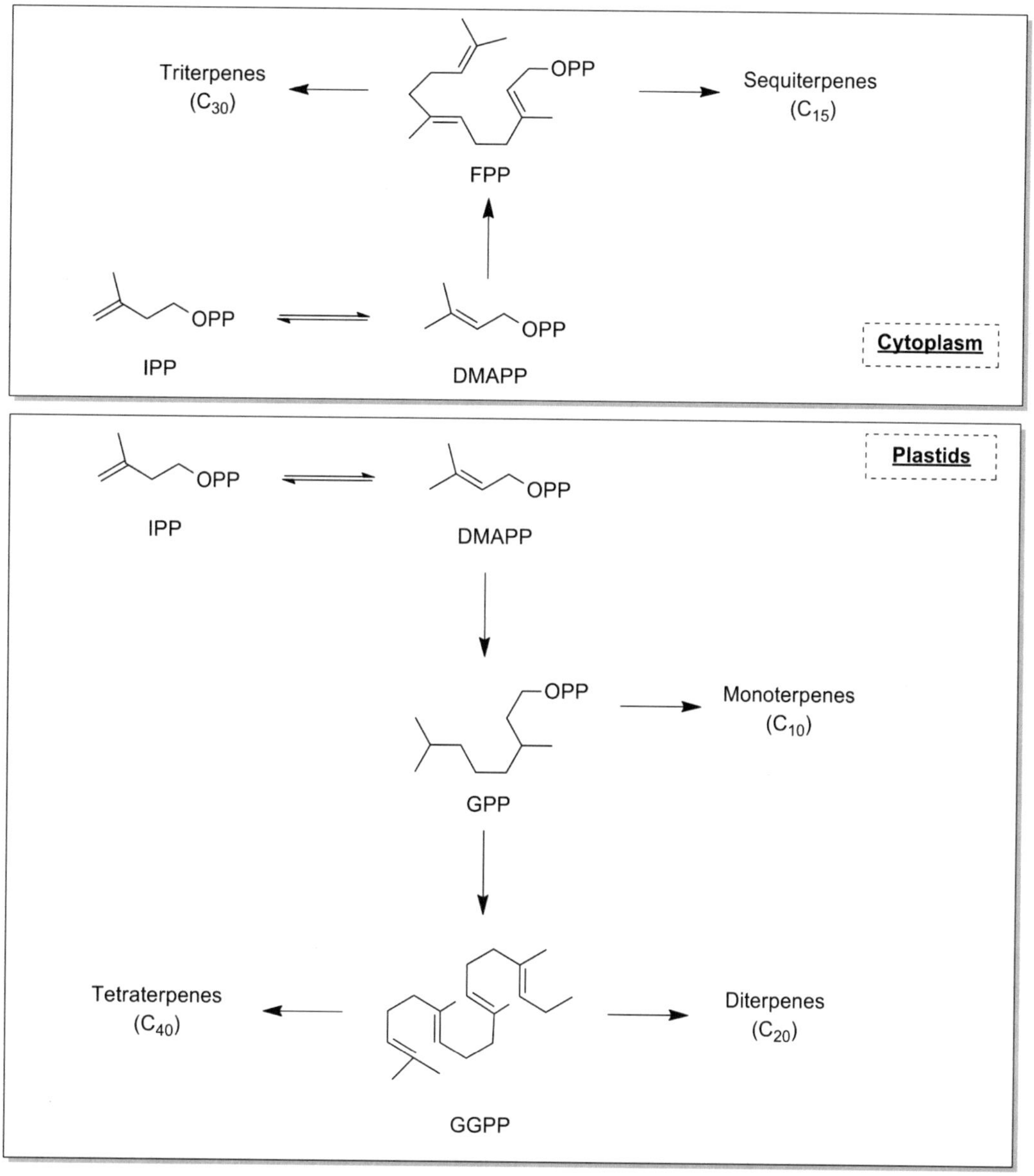

Fig. 4.1 Biosynthesis of terpenoids (Tholl 2006)

followed by rearrangement (Tholl 2006). Class I TPS are ionization-initiated and the active sites are located within α domain, which adopts a common α bundle fold. The aspartate-rich residues (DDXXD) bind a magnesium cluster that triggers the departure of the substrate diphosphate leaving group and concurrently initiates the cyclization and rearrangement reaction. Class II TPS are the protonation-initiated and the corresponding active sites reside between β/γ domains, both of which exhibits an α-barrel fold, in which a DXDD motif in the β domain provides the proton donor that triggers initial carbocation formation (Kampranis et al. 2007).

There has been an extensive study on the analysis of plant TPS with respect to understanding function and evolution of the TPS family (Chen et al. 2011; Bohlmann et al. 1998).

Table 4.1 Details of TPS subfamily and lineage-specific distribution

Subfamily	Groups	Functions	Distribution
TPS-a	TPS-a-1	SesquiTPS	Dicos
	Tps-a-2	SesquiTPS	Monocots
TPS-b		MonoTPS, IspS	Angiosperms
TPS-c		CPS/KS, CPS, other DiTPS	Land plants
TPS-d	TPS-d-1	Primarily MonoTPS	Gymnosperms
	TPS-d-2	SesquiTPS	Gymnosperms
	TPS-d-3	Primarily DiTPS, SesquiTPS	Gymnosperms
TPS-e/f		DiTPS, monoTPS, SesquiTPS	Vascular plants
TPS-g		MonoTPS, SesquiTPS, DiTPS	Angiosperms
TPS-h		Putative bifunctional DiTPS	*S. moellendorffii*

Many times, products of TPS are converted into bioactive natural products, terpenoids, in which the hydrocarbon backbones are decorated with oxygen by cytochromes P450. The primary drivers of terpene diversification are terpenoid synthases and cytochrome P450, which synthesize and modify terpene scaffolds.

4.1.3 Cytochrome P450

Cytochrome P450 monooxygenases (CYP450) are heme-containing enzymes catalyze a wide variety of monooxygenation reactions in primary and secondary metabolism in plants. The largest group of P450s is found in plants is one reason for the immense amount of structurally diverse natural products in this kingdom. CYP450s catalyze a large variety of reactions like aliphatic and aromatic hydroxylations, N-, O-, and S-dealkylation, oxidative deamination, sulfoxide formation or N-oxidation (Schuler 1996).

Phylogenetic analyses, based on translated raw DNA sequence data, have spaced the CYP450s into 10 clans that include 59 families and an extensive number of subfamilies. The sequence identity between distantly related CYP450s is very low, but a few absolutely conserved sequence motifs (such as the WxxxR motif, the GxE/DTT/S motif, the ExxR motif, the PxxFxPE/DRF motif and the PFxxGxRxCxG/A motif) are observed to be important substrate recognition sites (Gotoh 1992). Based on phylogenetic classification, the type of CYP450s is distinguished with its family number and subfamily letter, with shared numbers used for orthologs in different species and unique numbers used for paralogs in the same species. The nomenclature for P450s comprises of the CYP prefix, followed by an Arabic numeral which designates the family shares more than 40% identity with one or more previously confirmed P450s. Furthermore, if two CYP450s share more than 55% identity, they are considered to be members of the same subfamilies, designated with a capital letter and an Arabic numeral designates the individual gene (Nebert et al. 1991; Chapple 1998). The characterization and functions of CYP450s have been studied systematically in many organisms (Nelson 2009) http://drnelson.uthsc.edu/CytochromeP450.html.

4.1.4 Rhomboid Proteases

Rhomboids are transmembrane serine proteases, with a catalytic dyad of serine and histidine, conserved across the three kingdoms of life (bacteria, archaea, and eukaryotes). A member of the family Rhomboid 1 is found to be conserved throughout the kingdoms of life, suggesting an important role of the family. The proteins were first discovered in a mutational screen experiment in *Drosophila* by Christiane Nüsslein--Volhard and Eric Wieschaus (Mayer and Nüsslein-Volhard 1988). The protein was named after the rhomboid-shaped heads in mutants. In *Drosophila*, rhomboid protein (Rhomboid 1) controls the activation of the ligand for EGF receptor, by cleaving the transmembrane region of Spitz in the Golgi complex and releases the active ligand (Urban et al. 2001). Subsequently, new rhomboid sequences were identified in *Drosophila* and the human ortholog was also found to activate EGF signaling with a similar mode of action (Wasserman et al. 2000). Analysis of rhomboid substrates reveals that they are transmembrane helices with a partially disordered conformation (Fig. 4.2). Substitution of the residues with helix-breakers and β-branched residues enhances cleavage (Urban and Freeman 2003).

In addition to animals, rhomboids have also been studied in microorganisms and plants. Rhomboids in *Toxoplasma gondii* cleave the transmembrane micronemal (MIC) proteins, which are important for host invasion (Urban and Freeman 2003). Rhomboid protease AarA (in *Providencia stuartii*) activated quorum sensing by cleaving the N-terminal extension of the transmembrane protein TatA, activating its transport function (Stevenson et al. 2007).

Eighteen rhomboid sequences have been identified in the model plant *Arabidopsis thaliana* (Koonin et al. 2003; García-Lorenzo et al. 2006). *Oryza sativa*, *Populus trichocarpa* and *Physcomitrella patens* have 18, 16, and 13 sequences, respectively (Tripathi and Sowdhamini 2006; Lemberg and Freeman 2007; Kmiec-Wisniewska et al. 2008). Rhomboid sequences from some plant, invertebrates and vertebrates, were clustered into different clades and specific conserved regions identified using MEME were analyzed from the clades (Li et al. 2015). The study was limited to a few plant species, and a detailed analysis of conservation functionally important residues in different clades has not been looked at.

Although rhomboids have been identified and studied from many plants, the functions of these families and the substrates that they act on are

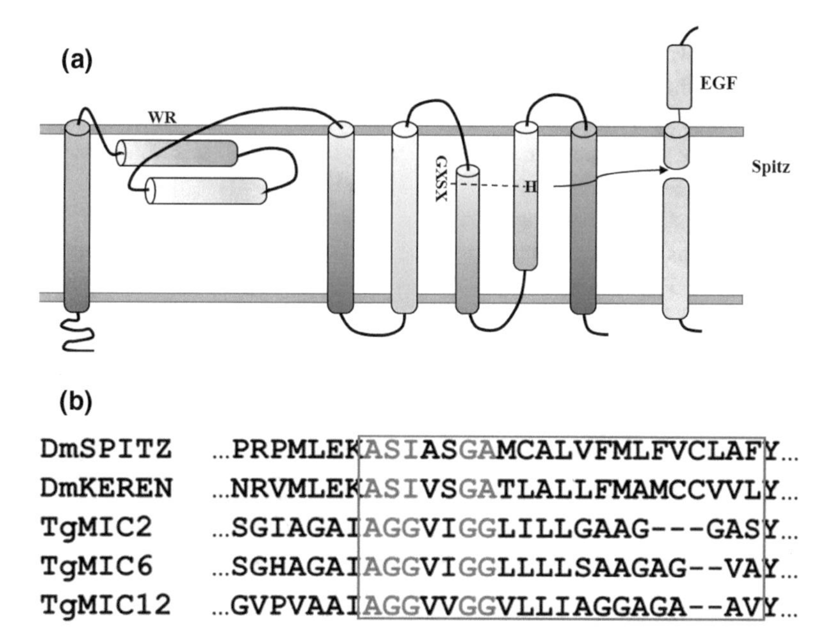

Fig. 4.2 Rhomboid structure and sequence features of substrates. **a** The proposed mechanism of action of Rhomboid 1 from *D. melanogaster* on Spitz ligand (adapted from Urban et al. 2001). **b** Alignment of the transmembrane helices (indicated by a box) of rhomboid substrates Spitz and Keren from *Drosophila* and TgMIC2, TgMIC6, and TgMIC12 from *T. gondii*. The residues corresponding to the substrate motif and GA motif of Spitz in the alignment have been highlighted with an outline (adapted from Urban and Freeman 2003)

still unknown (Knopf and Adam 2012). Recently, an Arabidopsis rhomboid, AtRBL10, was found to effect floral morphology (Thompson et al. 2012). In a previous study (Lemberg and Freeman 2007), plant rhomboids were seen to fall into four main classes, secretase B, mixed secretases, PARL and mixed inactive (Table 4.2). This classification suggested four divergently evolved lineages of rhomboids. The former two classes were active serine proteases localized in secretory pathway. The latter two classes were inactive enzymes with PARL being located in the mitochondrial membrane.

There have been attempts to computationally study the different families in specific plants to understand the tissue-expression patterns and function. Identifying the rhomboid sequences and analyzing the cellular localization, functionally important residues and associated protein domain families would help us understand the different classes of rhomboids in plants and the putative functions.

Recently, we sequenced and annotated the *Ocimum tenuiflorum* (Ote) genome (Upadhyay et al. 2015). Hence, it was possible to examine the above mentioned three important enzyme families at the whole genome level. This involved identification of the respective enzyme entries, through sequence-based approaches and further validation by structural features and phylogenetic relationships. These results can be used for functional validation of the Tulsi rhomboid genes and increase our understanding of the roles of plant rhomboids.

4.2 Materials and Methods

4.2.1 Sequence Search Methodology for Collecting TPS and CYP450 in the *Ocimum tenuiflorum* (Ote) Proteome

Annotated homologous sequences of ***TPS*** were collected from previously published reports (Chen et al. 2011) and along with reviewed entries from Swissprot database. These sequences were aligned using MUSCLE tool and an HMM profile was built using HMMER package (Baldi et al. 1994). The HMM profile of classical members was used as start point for searching homologous sequences in TulsiDB (http://caps.ncbs.res.in/Ote/) using BLASTp (Altschul et al. 1990). The resulting hits were then analyzed for domain assignment using Pfam to reduce false positives. Additionally, sequences less than 300 amino acid length were removed to avoid accumulation of incomplete sequences. Furthermore, redundant sequences sharing more than 95% mutual sequence identity were removed using CD-HIT.

The collection of annotated entries of CYP450 sequences involved more careful method considering the huge diversity and

Table 4.2 Details of rhomboid class distribution in plants from Lemberg and Freeman (2007)

Class	Functions	Distribution
Secretase B	Located in secretory pathway, 6 + 1 TMD topology, WR motif not absolutely conserved, GxSxxxF catalytic motif, predicted to be the ancestral precursor, cleave type I membrane proteins (N_{out}/C_{in})	*A. thaliana, O. sativa*
Mixed other secretases	May represent evolutionary intermediates between an ancestral rhomboid (of the B type) and the potentially more recently evolved rhomboids of the higher eukaryotes (Secretase A type)	*A. thaliana, O. sativa*
PARL (presenilin-associated rhomboid-like)	1 + 6 TMD topology, predicted mitochondrial localization, WR motif within the predicted TMD2, substrates have an N_{in}/C_{out} topology	*A. thaliana, O. sativa*
Mixed inactive homologs	Inactive rhomboid-like genes, relatively recent mutations of active rhomboids, not found in other plants, presumably derived from Arabidopsis PARL	*A. thaliana*

abundance of this enzyme family. Sequences from *Mimulus guttatus (Mgu), Solanum lycopersicum,* and *A. thaliana* were collected from CYP450 database (Nelson 2009) and used as references for searching homologs in Ote database. *M. guttatus* was chosen as it shares close homology with Ote, while *S. lycopersicum* and *A. thaliana* were chosen as their genomes are well-annotated. In order to increase the sensitivity of sequence search, multiple HMM profiles were built for major cytochrome families and were used as start points during sequence search in Ote genome. Also, individual BLAST searches, with each reference sequence, were performed to avoid missing any Ote entries. Redundant hits from all of the sequence search runs were removed using CD-HIT clustering. The non-redundant hits were queried against Pfam database to remove false positives. Figure 4.3 explains the sequence search workflow followed for identifying TPS and CYP450 sequences in Ote genome.

4.2.2 Structural Alignment to Compare the Functional Important Residues

The Terpene synthase/cyclases can be broadly divided into two superfamilies depending on the location of active site. The active site is characterized by the presence of either a DDXXDD or DDXD motif in the active site. Based on the relative position of this motif, terpene cyclase can be either classified into alpha superfamily containing a C-terminal active site (SCOP Superfamily ID: 48576) beta and gamma superfamily where the motif is present in the (SCOP Superfamily ID: 48239). For the present study, 739 PDB entries from both the superfamilies were collected SCOP 2.05 (Fox et al. 2014) database. These hits were then reduced to total of 239 PDBs by removing redundant entries using CD-HIT clustering. This protocol (Fig. 4.4) was implemented for each of the families to generate structure-based alignment.

4.2.3 Phylogenetic Analysis of TPS and CYP450

The hits from sequence searches were aligned using MUSCLE program, a multiple alignment tool. Further, the alignment was processed for removal of spurious sequences or poorly aligned regions using the trimAL tool. The processed alignment was used for phylogeny construction by Randomized Axelerated Maximum Likelihood (RAxML) with 1000 bootstrap replications. Sequences sharing more than 90% mutual sequence identity were removed using CD-HIT

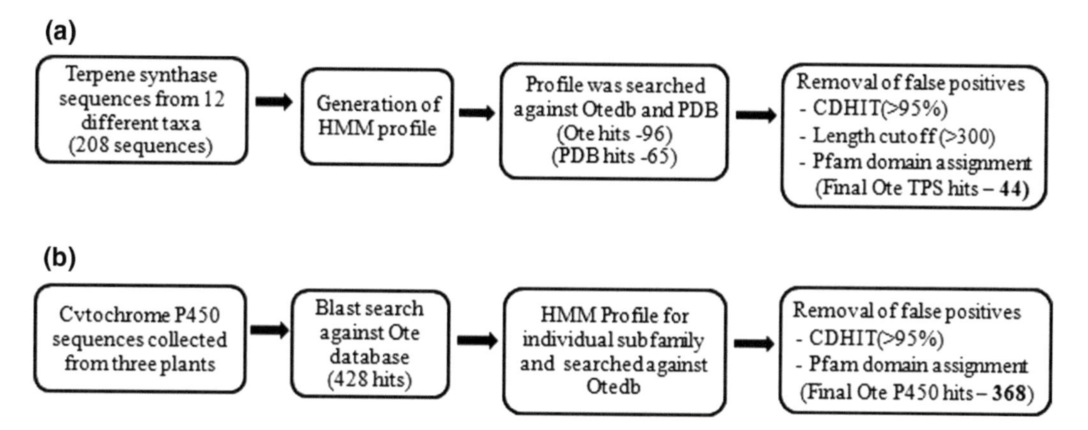

Fig. 4.3 Flowchart for sequence-based searches for both TPS and cytochrome p450

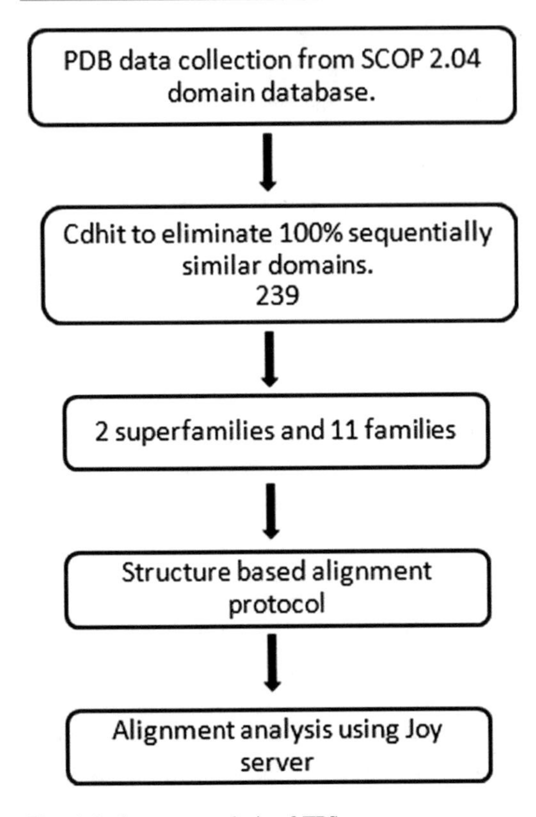

Fig. 4.4 Structure analysis of TPS sequences

to reduce the computational time for construction of phylogenetic tree. The resulting phylogeny was viewed using a graphical viewer tool Figtree (Rambaut 2009).

4.2.4 Identification of Rhomboid Proteases in the Ote Proteome

Using 18 known rhomboid sequences from *A. thaliana* and 18 from *O. sativa japonica* (Tripathi and Sowdhamini 2006) as queries, a multi-fold sequence search approach was employed to search for rhomboids in Ote (Krishna Tulsi subtype) from the draft genome from NCBS (Upadhyay et al. 2015). Multiple sequence search methods were used to enhance the number of hits. Since the sequence identity between the rhomboid sequences is very low, sensitive remote homology prediction methods were used. The tools used were Context-Specific BLAST (CS-BLAST) (Biegert and Soding 2009), Pattern-Hit Initiated BLAST (PHI-BLAST) (Zhang et al. 1998), TBLASTN (Gertz et al. 2006) and BLASTX (Camacho et al. 2009). Profile-based methods like Reverse-PSI BLAST (RPS-BLAST) using Conserved Domain Database (CDD) (Marchler-Bauer et al. 2002) profiles, HMMSCAN and HMMSEARCH (using Pfam (Sonnhammer et al. 1998) HMMs) were also used. BLAST+ version 2.2.31, legacy BLAST version 2.2.24, and Hmmer version 3.1.b1 were used for the sequence searches. A set of unique hits was constructed from the above sequence search pipeline. The unique sequences, identified from the sequence search method, were further screened for criteria like the presence of 6–7 transmembrane helices and functionally important residues. An all-against-all BLAST was carried out to identify the orthologs and paralogs among the proteins from all three species. The sequence search workflow has been described in detail in Fig. 4.5.

4.2.5 Identification of Transmembrane Helices and Subcellular Locations of the Rhomboid Proteins

The rhomboid structural domain has six transmembrane helices. Some plant rhomboid sequences are predicted to have an extra transmembrane helix, either at the N-terminus or the C-terminus. Tools like Psipred (Jones 1999) were used for identifying secondary structures and Phobius (Käll et al. 2004) and TMPred (Hofmann 1993) for identifying transmembrane helices. Literature on Arabidopsis orthologs was used to predict cellular localization of Tulsi rhomboids. Target-P tool was used to predict cellular localization of the remaining rhomboids.

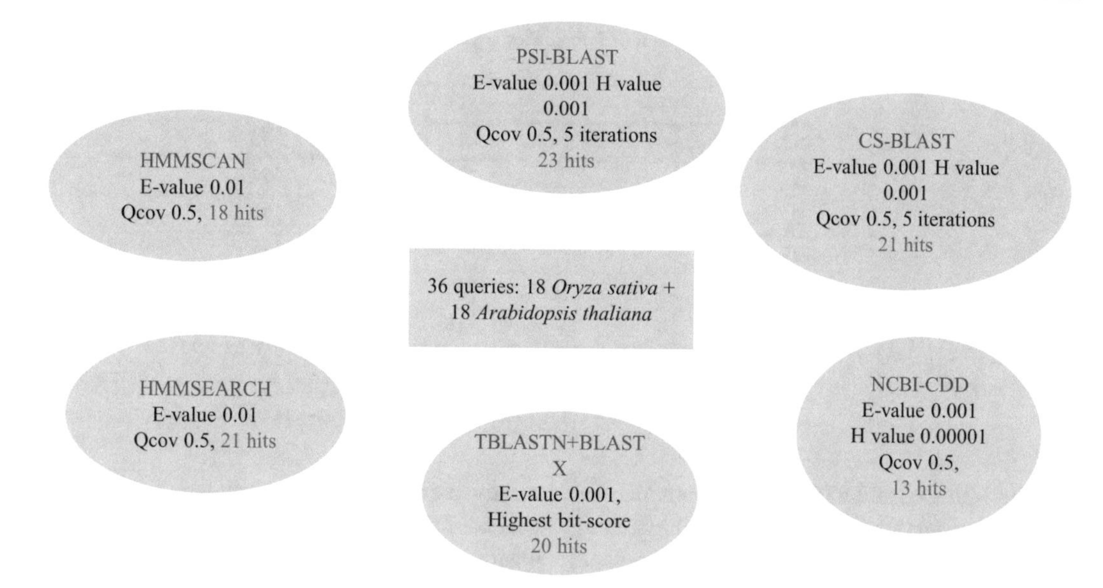

Fig. 4.5 Different methods used for sequence search with the tool version and parameters used have been shown in the figure. A total of 36 rhomboid sequences from *Arabidopsis* and rice were used as queries to search the Tulsi proteome

4.2.6 Multiple Sequence Alignment, Phylogeny and Identification of Variants in Functionally Important Residues

HMMalign (Eddy 1998) from Hmmer 3.1b1 was used for aligning the identified rhomboid sequences with the query sequences from *Arabidopsis* and rice, guided by Pfam (v 28) rhomboid family HMM. The sequence domain starts a few residues before the first loop, and the first transmembrane helix is not a part of the domain. Serine and histidine residue are involved in peptide cleavage (Lemberg et al. 2005). The conservation of functionally important residues 'WR' in the first loop, 'GXSX' in the fourth helix, and 'H' in the sixth helix was studied. Neighbor joining and maximum parsimony from Phylip (Felsenstein 1981) package were used for the construction of the trees. The sequences were classified into different classes based on the variation in the functionally important residues.

4.2.7 Prediction of Domain Architecture

CDD and HMM profile matching were carried using RPS-BLAST and HMMSCAN to identify the co-occurring domains with rhomboids. The results of the tool which predicted more number of domains and rhomboid domain length were recorded.

4.3 Results and Discussion

4.3.1 Sequence and Structure-Based Analysis of Newly Identified Ote (TPS)

The identified TPS sequences from Ote genome were clustered well in all subfamilies except for D and H subfamily (Table 4.3). D subfamily is known to be present only in Gymnosperms, and the H subfamily is known to exist only in *Selaginella moellendorffii* (Table 4.1). These TPS are

Table 4.3 Distribution of Ote TPS members in different subfamilies

Subfamily	No. of Pdb entries	No. of Ote entries	No. of classical members
A1	11	11	64
A2	–	1	33
B	–	7	22
C	1	5	11
D1, D2, D3	–	–	16
E/F	–	5	23
G	4	15	27
H	–	–	6
Total	16	44	202

putative bifunctional DiTPS. A maximum number of Ote hits were identified in G subfamily hinting at the excessive amount of geranyl diphosphate present in the cytosol, as this the major precursor in most reactions synthesized by enzymes of this family. This family also correlates with the known fact of active volatiles produced by the basil plant. A1 subfamily enzyme shows high sequence (>50%) identity with major sesquiterpene synthase and majority fall under cadinene synthase type of terpenes, including gamma-cadinene, muurolene, amorphene which are major enzymes in essential oils producing plants.

To further validate the hits from structural perspective, the G subfamily Ote sequences were aligned with the corresponding known PDB entries. The presence of characteristic DDXXD motif, which is important for alpha domain functioning, was clearly observed. A similar kind of alignment was performed for other subfamilies with the respective PDB sequences and similar trend has been observed.

The structural alignment of families showed conservation of certain sets of residues which are functionally important for enzyme action (Fig. 4.6). These tend to be class-specific conserved residues (observed within the family) and are listed in Table 4.4.

Alignment at the superfamily level revealed new patterns, such as the amino acid change from NSE (NDXXSXXXE) to DTE (DDXXTXXXE) motif, to show clear evolutionary change of functionally important residues (Figs. 4.7, 4.8, 4.9). The superfamily 48586 mostly comprises of bacterial and fungal TPS, whereas as we move to 48583 it comprises of plant-based TPS. The clustering seen in the phylogeny of 272 reported entries (Fig. 4.10) corresponds well with reported clustering patterns. Such clustering shows that all the new sequences fit properly with the known classification of TPS. All these subfamilies have been known to show specific characterized trait which can be summarized in Table 4.4 (Chen et al. 2011).

The presence of conserved motifs in the newly identified sequences helps to accept them as true hits. The new sequences have shown characteristic DDXXD and DXDD motifs. For instance, sequences from Tulsi, which cluster in the C subfamily, show clear the presence of DXDD motif, which is found in β-γ active domains. Similar patterns were found in all the new sequences.

4.3.2 Sequence and Structure Analysis of CYP450

CYP450 represents one of the largest classes of proteins, containing more than 50 subfamilies. In our sequence search and phylogenetic analysis of CYP450 sequences in Ote, CYP71 subfamily was found to be the largest subfamily, which is in agreement with previous observations (Nelson 2009). Nearly 40% of all the hits belonged to five subfamilies (CYP76, CYP716, CYP72, CYP736 along with CYP71), which are known to be

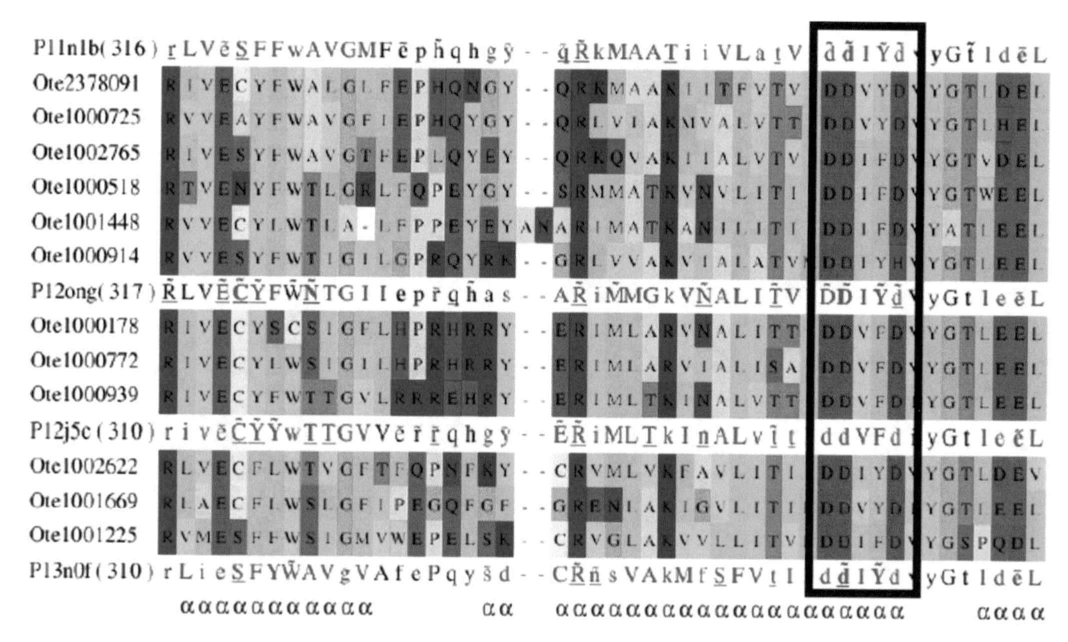

Fig. 4.6 Structural alignment of G subfamily Ote sequences with known structures

Table 4.4 Details of residue conservation across TPS family and superfamily

Superfamily	Family[a]	Conserved residues[b]
48239	48240 (Terpenoid cyclase N-terminal domain)	E29, L38, R43, L47, Q54, L55, I59, I60, L66, (EL) 77, L79, F122, FKASLA(133-138), G143, LYEAS(146-150)
	48243 (Terpene synthases)	WDT(A/G) 104-107, L135, GG(157-158), T172, L227, T250, (L/I) 277, 285-290, YGT(303-305), DGGW(346-349), Y362, TXWAM (373-376), GV(398-399)
	227201	ID(76-77), RLGI(82-85), F90, FRL(L/M)R(127-131), F144, G150, Y168, F187, L206, I/V(210), AL(213-214)
48576	48577 (Isoprenyl diphosphate synthases)	E94, L100, DDXX(XXX)D(103-110), RG (115-116), K207, DDXXD(250-254)
	48583 (Terpenoid cyclase C-terminal domain)	N2, Q17, QE(22-23), (V/L)SRWW(27-31), F36, L40, V47, Q62, R67, M69, DDXXD(81-85), GT(88-89), ELE(92-4), D98, QR (100-101), RLP(129-31), YM(113-114), SY(118-119), D128, 165-169, EYL(172-174), A176, DD(226-227), C245, EA (256-257), AWKD(269-272), NL(293-294), IL(323-324)
	48586 (Aristolochene/pentalenene synthase)	YFP(55-58), DDXXD(78-82), D(L/I)122-23, TD(151-152), QYL (168-170), R172

[a]Some families have been removed due to redundant entries
[b]Positions have been labeled as per the alignment file and not actual residue position

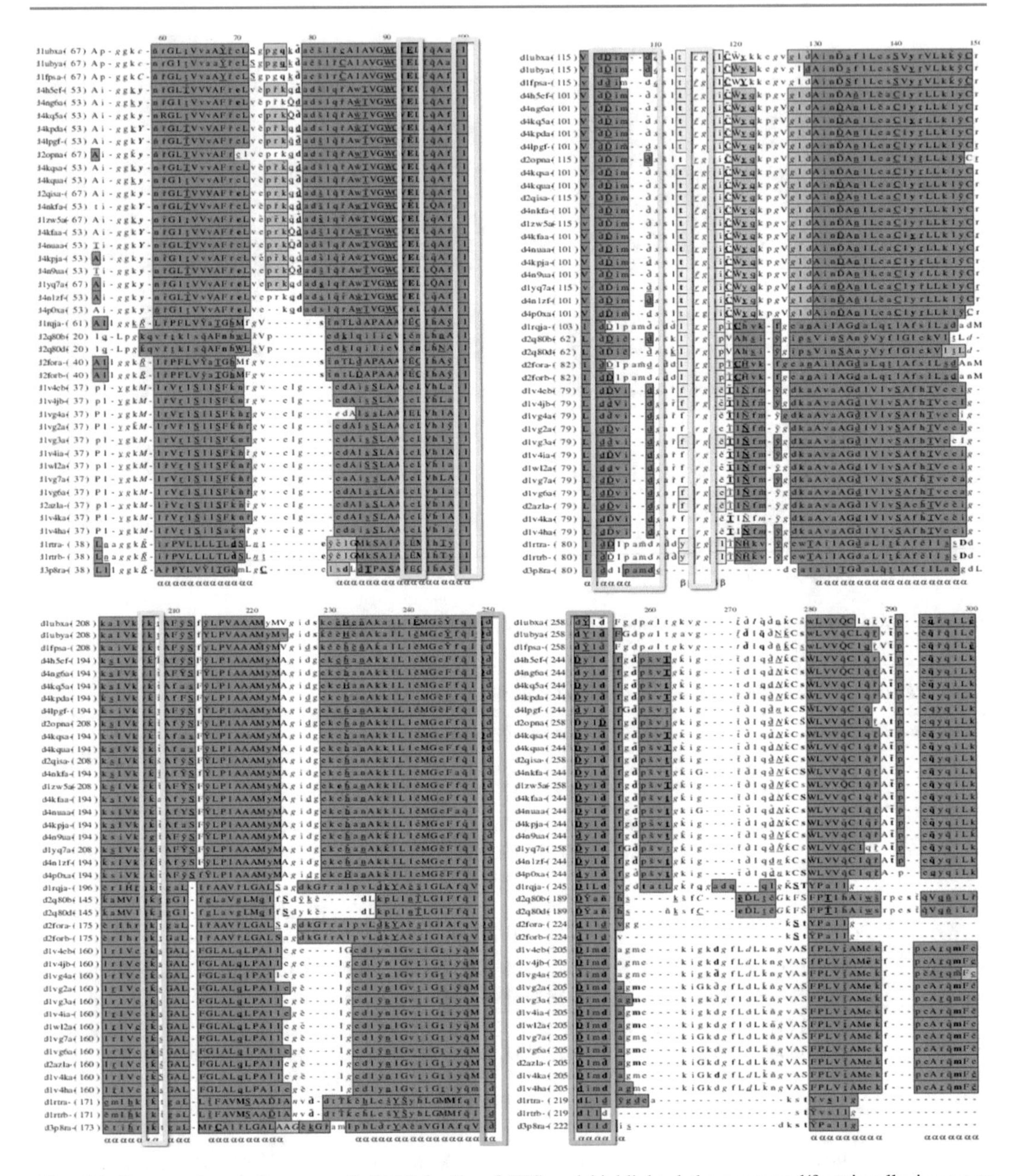

Fig. 4.7 Structure-based alignment of 48577 family of TPS and highlighted the conserved/functionally important residues on the alignment

present in large numbers (Fig. 4.11). A similar trend has been observed in previous analysis of large model organisms (Boutanaev et al. 2015).

The diversity in sequence space in CYP450 is mainly attributed to its various substrates. It has been proposed earlier (Gotoh 2012) that there are four sequence motifs (based on the nature of substrates), which are characteristic signatures of CYP450 family. In the present study, we have demonstrated usage of such sequence motifs in one of the subfamily CYP71, since similar analysis can be used a powerful tool in identification and classification of CYP450 family and its members. These regions for P450 71 subfamily are:

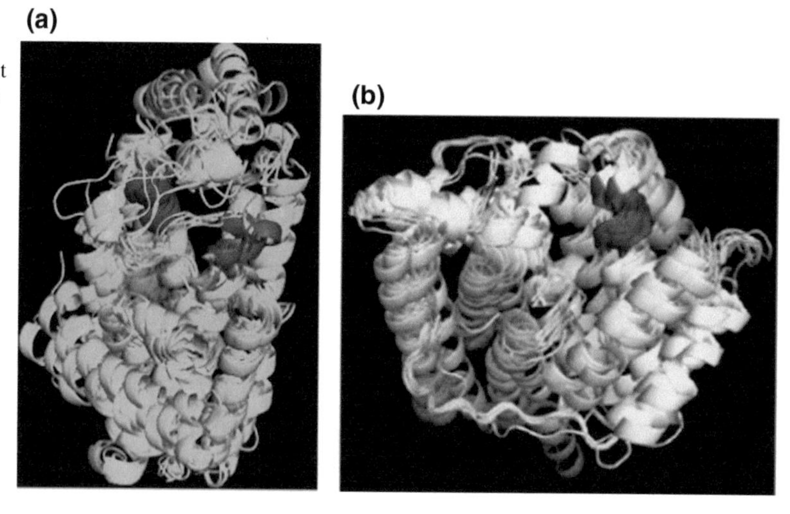

Fig. 4.8 **a** Alignment of isoprenyl diphosphate synthases family, **b** alignment of Aristolochene/pentalenene synthase red marked areas shows the conservation of DDxxD motif which stabilizes the Mg2+ ion cluster in the active site

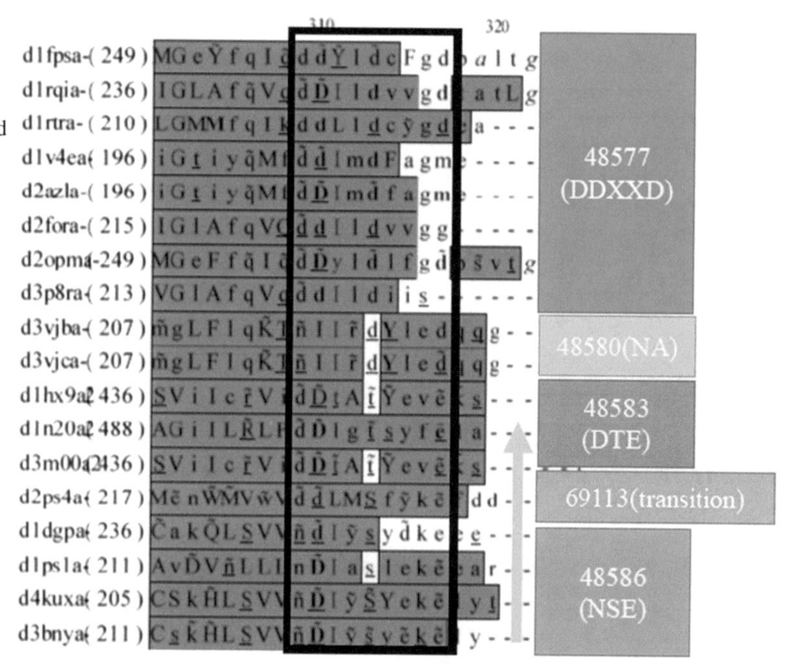

Fig. 4.9 Structure-based sequence alignment of TPS superfamily members of known structure. SCOP superfamily codes are marked on the right

1. SRS 6: Its site is located near C-terminus after PFXXGXRXC motif (around 30aa).
2. SRS 5: It is rich in 'Proline' and is located just adjacent to (after) EXXR motif and spans for about 15aa.
3. SRS 4: Around 50 amino acids, toward N-terminal from SRS 5, are a conserved GXXTS type motif and SRS4 spans upstream of it for around 10 amino acids
4. SRS1: At around 150th (±50) sequence position, there is a conserved WXX. (X/R) site and few tyrosines (Y) are observed to be conserved 20 amino acids before the SRS1 region.

All of the four well-defined SRS sequence motifs were well conserved in CYP71 subfamily of sequences from Tulsi and the alignment is shown in Fig. 4.12.

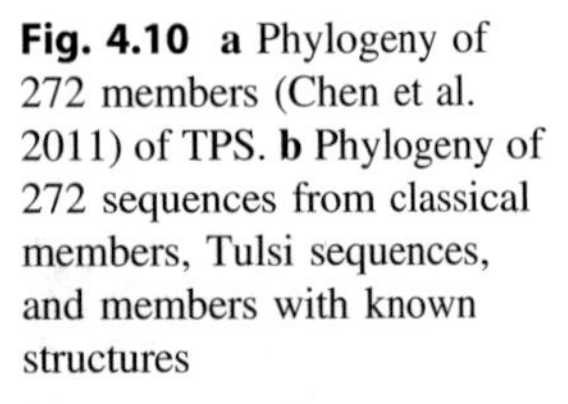

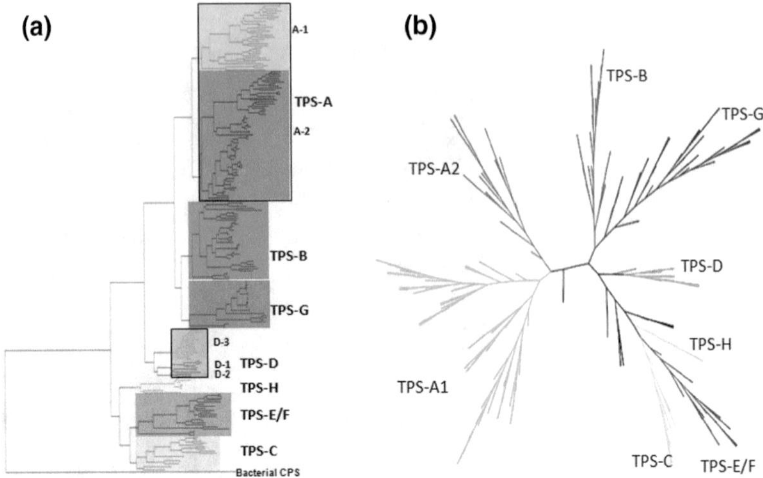

Fig. 4.10 **a** Phylogeny of 272 members (Chen et al. 2011) of TPS. **b** Phylogeny of 272 sequences from classical members, Tulsi sequences, and members with known structures

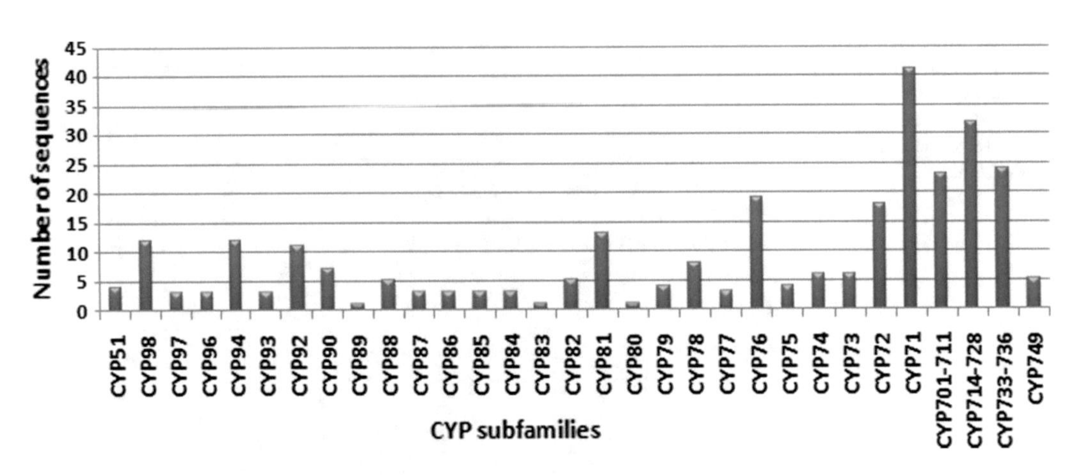

Fig. 4.11 Distribution of Ote CYP450 sequences based on their subfamilies

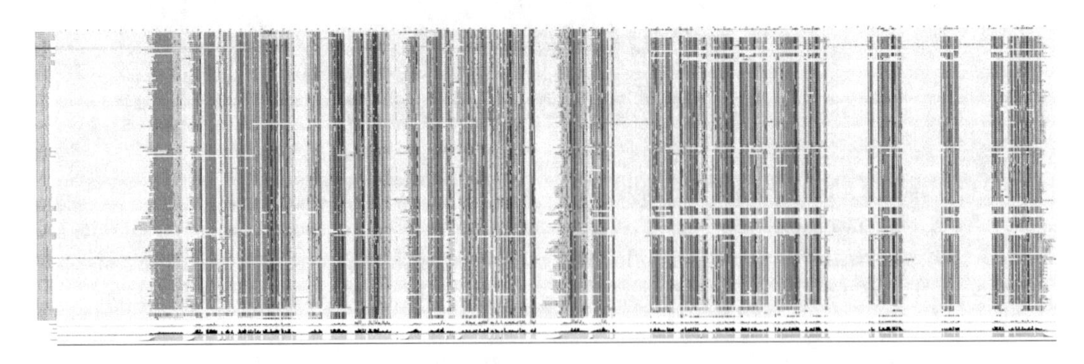

Fig. 4.12 Mapping of SRS residues on cytochrome P450 subfamily 71

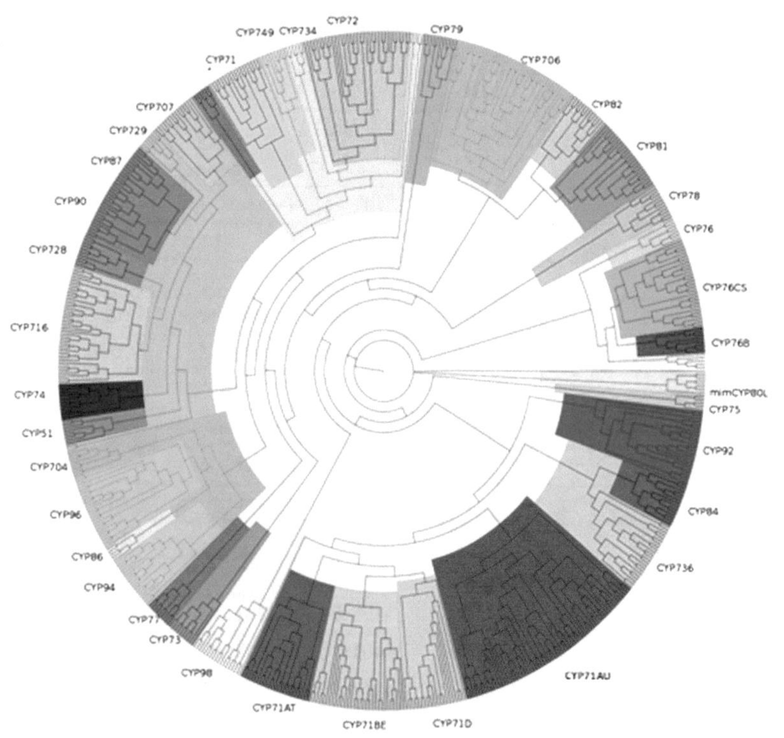

Fig. 4.13 Phylogenetic tree of Ote hits and monkey flower P450s

Phylogenetic analysis of identified Ote CYP450 sequences with those from monkey flower *[M. guttatus (Mgu)], S. lycopersicum* and *A. thaliana* showed that Ote P450 sequences cluster very well with *M. guttatus (Mgu)* than other plant genome sequences (Fig. 4.13).

4.3.3 Domain Architecture Analysis of Ote Rhomboid Hits

Twenty-two rhomboid sequences were obtained after the screening. For ease of analysis, the sequences were named Ote1 to Ote22. A previous study by Freeman and coworkers in 2009 had classified rhomboid sequences from eukaryotic model organisms like yeast, *Drosophila*, human, and plants (*A. thaliana* and *O. sativa*) into the following functional groups—PARL-type, secretase, mixed inactive types, and i-Rhoms. Plant rhomboids were found to be of the former three types. Most of the Tulsi rhomboids were found to be from the mixed other secretase class, followed by Secretase B (3 sequences) and one sequence from PARL class. The list of rhomboid hits in Ote, the orthologs in Arabidopsis and rice have been provided in Table 4.5.

Most of the hits are found to occur as single domains. A few of them are associated with other domains like ribonuclease II catalytic domain (RNB), virus attachment protein p12, ubiquitin-associated domain (UBA), C-terminal domain of the auxin-interacting protein family (Aida_C2), cytochrome C biogenesis protein (CcmH), domain of unknown function (DUF) 1421, positive regulator of the transcription factor sigma (E) (RseC), and ribosomal protein S7 (RpsG). The combined domain predictions from Pfam and CDD have been provided in Table 4.6.

The hits have at least one of these Pfam families—rhomboid, DER1 and DUF 1751 (except Ote21 which does not have any predicted Pfam domain). Since Ote21 matched the CDD profile for DUF 1751, we retained the sequence for further analysis.

Table 4.5 Rhomboid classes in *O. tenuiflorum* and the corresponding orthologs in *Arabidopsis* and rice have been indicated. The classes were assigned from the Arabidopsis orthologs

O. tenuiflorum protein id	Hit code	Assigned Rhomboid class	*Arabidopsis* ortholog id	Rice ortholog id
237723150011	Ote1	Secretase B	At3g58460	Os03g44830
100106620051	Ote2	Mixed other secretase	At1g63120	Os11g47840
100110990031	Ote3	Mixed other secretase	At1g63120	Os11g47840
100078830091	Ote4	PARL	At1g18600	Os01g55740
100251520091	Ote5		–	–
100018530131	Ote6	Mixed other secretase	At5g38510	Os11g47840
100205230051	Ote7	Mixed other secretase	At1g63120	Os09g35730
100251720021	Ote8		–	–
100197820111	Ote9	Mixed other secretase	At2g29050	Os08g43320
100049210061	Ote10	Mixed other secretase	At1g25290	Os09g28100
100104910091	Ote11	Mixed other secretase	At5g07250	Os08g43320
100213940171	Ote12		At1g63120	Os10g37760
100015610031	Ote13	Secretase B	At3g59520	Os01g67040
100078690041	Ote14	Mixed other secretase	At5g07250	Os09g35730
100038730091	Ote15	Mixed other secretase	At5g07250	Os11g47840
100071920081	Ote16	Mixed other secretase	At5g25752	Os05g13370
100243700071	Ote17		–	–
100172030131	Ote18		At3g56740	Os01g16330
100231260071	Ote19	Secretase B	At3g17611	Os01g18100
100149160031	Ote20		At5g07250	Os07g46170
100132970031	Ote21		At5g07250	Os03g24390
100278350051	Ote22	Mixed other secretase	At1g77860	Os11g47840

Few hits could be related to Pfam domain rhomboid and distantly related families of rhomboids like DUF1761 and DER1. Since these two families are closely related to rhomboid family, the hits were retained.

4.3.4 Predicting the Cellular Localization, Function, and TM-Helices of Ote Rhomboid Hits

Most of the rhomboid hits were predicted using Target-P to be a part of secretory pathway, followed by Mitochondria (Table 4.7). This is consistent with a previous analysis that secretory type of rhomboids is more common in plants (Lemberg and Freeman 2007). Orthologs of Ote7, Ote9 and Ote12 have been shown to occur in Golgi and are expressed in all tissues by Kanaoka and coworkers. Ote4 is the ortholog of Arabidopsis PARL rhomboid which occurs in mitochondria and is predicted to be inactive (Kmiec-Wisniewska et al. 2008). Ote10 and Ote16 orthologs in Arabidopsis are present in plastids and are involved in floral morphology and senescence (Knopf and Adam 2012; Thompson et al. 2012). One of the rhomboid orthologs was found to be overexpressed under heat shock (Ote14). Almost all the predictions

Table 4.6 DA obtained from Pfam and CDD for the hits in *O. tenuiflorum*. The domain coordinates and sequence length (in bracket) have been provided

Hit code	CDD domain architecture	Domain coordinates
Ote1	Rhomboid superfamily	84-241(347)
Ote2	Rhomboid superfamily ~ CcmH superfamily	92-236 ~ 55-285(357)
Ote3	Rhomboid superfamily ~ RNB superfamily	82-226 ~ 96-992(1101)
Ote4	Rhomboid superfamily	139-303(307)
Ote5	Rhomboid superfamily	12-211(242)
Ote6	Rhomboid superfamily	237-381(433)
Ote7	Rhomboid superfamily ~ P12	96-240 ~ 265-312(313)
Ote8	Rhomboid superfamily	11-201(304)
Ote9	Rhomboid superfamily	108-285(352)
Ote10	Rhomboid superfamily	108-252(279)
Ote11	Rhomboid superfamily	104-258(331)
Ote12	Rhomboid superfamily	103-252(273)
Ote13	Rhomboid superfamily	1-130(196)
Ote14	Rhomboid superfamily	110-254(328)
Ote15	Aida_C2 ~ Rhomboid superfamily	1-44 ~ 15-259(381)
Ote16	Rhomboid superfamily	119-277(282)
Ote17	DER1	11-199(237)
Ote18	Rhomboid superfamily ~ DUF1421 ~ UBA_like_SF superfamily	82-232 ~ 49-314 ~ 90-326(332)
Ote19	Rhomboid superfamily	67-221(302)
Ote20	DUF1751	38-148(341)
Ote21	DUF1751 ~ RseC_MucC Superfamily ~ RpsG	1-19 ~ 6-78 ~ 27-178
Ote22	Rhomboid superfamily	499-643(767)

agreed with the literature evidence for localization of the Arabidopsis ortholog. An exception was Ote19, predicted to be mitochondrial (low confidence), but the Arabidopsis ortholog was found in the secretory pathway.

Plant rhomboids have six to seven transmembrane helices. Since most of the transmembrane helix prediction methods have not been trained on plant transmembrane protein sequences, few rhomboid hits were predicted to have less than six TM-helices. Nevertheless, such sequences aligned well with the known rhomboid sequences.

4.3.5 Alignment to Identify Functionally Important Residues

The hits obtained from Ote proteome were aligned with the queries from the other plants. The residues WR in the second loop, the motif GXSX in the fourth helix and the residue H in the sixth helix are crucial for the functioning of the serine protease and were analyzed for conservation across the hits.

Table 4.7 Intracellular localization of rhomboid hits in *O. tenuiflorum* predicted using known information for Arabidopsis ortholog and Target-P. M refers to mitochondria, P to plastids, and C to chloroplasts in cellular localization

Hit number	Cellular localization	Literature on Arabidopsis ortholog
Ote1	Secretory pathway (Target-P)	Secretase-B type, higher levels compared to others (Baerenfaller et al. 2011)
Ote2	Other (Target-P)	Pollen elongation form Genevestigator tool prediction (Knopf and Adam 2012)
Ote3	Other (Target-P)	
Ote4	M (Kmiec-Wisniewska et al. 2008)	Mitochondrial PARL, forms a 6-TM topology, instead of the 1 + 6 topology characteristic of PARL
Ote5		
Ote6	M (Target-P)	Predicated as inactive
Ote7	Golgi (Kanaoka et al. 2005)	Expressed in all tissues
Ote8		
Ote9	Golgi (Kanaoka et al. 2005)	Expressed in all tissues
Ote10	P (Thompson et al. 2012) Chloroplast inner envelope	Mutants have reduced fertility and aberrant floral morphology (Thompson et al. 2012)
Ote11	Other (Target-P)	
Ote12	Golgi (Kanaoka et al. 2005)	
Ote13	Secretory pathway/other (Target-P)	Secretase-B type
Ote14	Other (Target-P)	Expressed in high levels during heat shock (Hruz et al. 2008)
Ote15	Other (Target-P)	
Ote16	P (Kmiec-Wisniewska et al. 2008), chloroplast inner envelope	Senescence from Genevestigator tool prediction (Knopf and Adam 2012)
Ote17		
Ote18	M (Target-P)	
Ote19	M (Target-P) (low confidence)	Secretase-B type
Ote20	Secretory pathway (Target-P)	
Ote21	Other (Target-P)	
Ote22	C (Target_P) (low confidence)	Predicted as inactive

The alignment also indicated the conservation of TM-helices and less conservation in the connecting regions between helices (Fig. 4.14). The Pfam domain starts from the second loop region and does not contain the first transmembrane helix.

The helices have been labeled and the motifs WR (second loop), GXSX (fourth helix), and H (sixth helix) have been indicated by a red box around the alignment region.

Some of the sequences from each plant have a ~30–40 residues insert region in the region connecting the first two TM-helices (Fig. 4.15). Secondary structure prediction using hhpred and Psipred predicts the region to be unstructured. Rhomboids with the insert were also present in other members of Viridiplantae (unpublished results). The conservation of the insert across the plant kingdom indicates that there might be a functional importance.

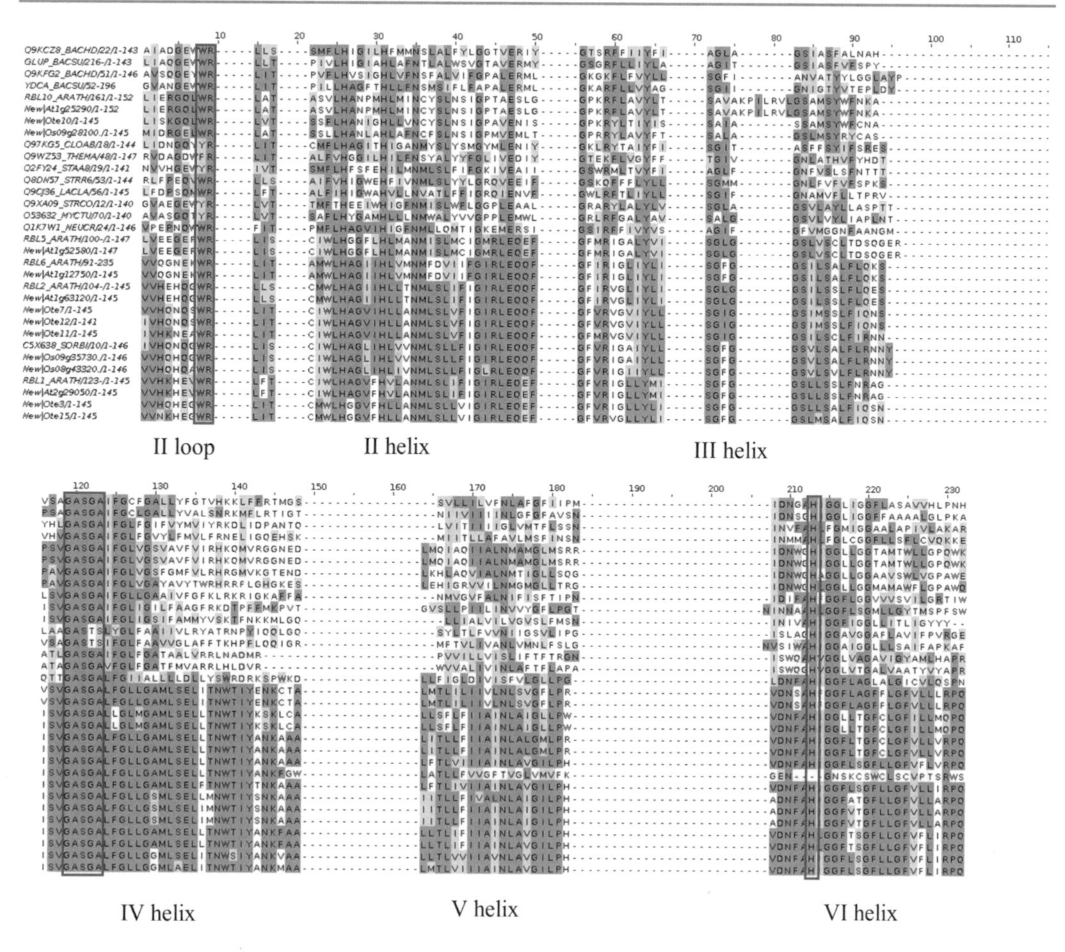

Fig. 4.14 An alignment of some of the rhomboids from Arabidopsis, rice, and Tulsi

4.3.6 Phylogeny and Orthology with Other Plant Hits

The consensus tree from NJ showed clustering of sequences according to similarity of functionally important residues (Fig. 4.16) and was used for further analysis. We have classified rhomboids into four functional classes based on phylogeny and conservation of residues:

4.3.6.1 Class I—All the Functional Residues Are Conserved

The sequences cluster with the Arabidopsis rhomboids from mixed other secretases as described in Lemberg and Freeman (2007). The sequences from Ote belonging to this class are Ote2, Ote3, Ote7, Ote9, Ote11, and Ote12. Our phylogeny results show these sequences cluster at the older branches which indicates that the class may be a primitive from which the other rhomboids have evolved.

4.3.6.2 Class II—Only GXSX Conserved

The WR motif is replaced with WX and H by any other residue. The sequences cluster with the known Secretase-type B and PARL proteins. PARL proteins are found in mitochondria and are inactive. The Ote sequences belonging to this class are Ote1, Ote4, and Ote19.

```
Ote2      -------------------MIHPVSVE--------------SPS---------------I
Ote15     MAAVSDTAPMLGQSRRSGNTIHPVSLE--------------SPSPAG-----------AG
Ote7      --------------------MRNGDLE------------KNRNRR--------YP-RYDM
Ote11     -------------------MMGNRDPE------------RGAGRKQS-----YYAGAYDM
At07250   ------------------MAVGDDDLENRMSAKDRGIGSRGGDRNRIGPPPLPVALSSST
Ote14     --------------------MRGQDIE-------------SKEEKGD-------HESAEA
At63120   --------------------MANRDVE------------RVGKKNRG--------ANNNY
Os35730   --------------------MASNGGE-------------EKSRVAAG----YGGGGYGY
Os43320   --------------------MASSSGE-------------GKGAGGAG----YQYAPYGG
Os47840   ---------------MKPSPAANLDVR------------VERPRPPP------VHPHRPG
                                    . .

Ote2      VYREVKHFKKWFPWLIPCFVVANILVFIITMYVNDCPKNSVS----------CSAGFLGR
Ote15     VYREVKHFKKWTPWLIPFFVVANVVMFVITMYVNNCPKNSIS----------CVARFLGR
Ote7      EYSE----SQWTSWLVPMIVVANVAVFVVIMYVNNCPKNHG---------GGCVAKFLGR
Ote11     DYSD----SQWTSWLIPMIVVANVAMFVVIMFVNDCPKNHDSF------RGDCVARFLGR
At07250   EFGDNALSSRWTSWLVPMFVVANVAVFVVAMFVNNCPNHFESHRL----RGHCVAKFLGR
Ote14     SYAAPAAERTWISWLIPVFVIANVAMFVIIMYFNNCPKRIRNRGFGYAGDDKCVARFLGR
At63120   FYEESSGETHWTSWLIPAIVVANLAVFIAVMFVNDCPKKITGP------NKECVARFLGR
Os35730   GGYEGRDDRKWWPWLVPTVIVACIAVFIVEMYVNNCPKHGSALGG-------CVAGFLRR
Os43320   SYYD--EERRWWPWLVPTVLVACIVVFLVEMFVNDCPRHGSPLRGE-----SCVAGFLHQ
Os47840   SLRARPYYRRWTPWIVAAIALSCVVVFLVSMYVNDCPRRNSGD---------CAAGFLGR
                    * .*::. . :: : :*:  *:.*:**..             * * ** :

Ote2      LSFQPFKENPLLGPSSSALEKMGALDVAKVVHGHEGWRLITCMWLHGGLFHLLANMLSLL
Ote15     FSFHPLKENPLLGPSSPALEKMGALDVDRVVNKHEGWRLITCMWLHGGVFHLLANMLSLL
Ote7      LSFQPLKENPLFGPSSSALQKLGGLEWNKVVHQNQSWRLITCIWLHAGVIHLLANMLSLV
Ote11     LSFQPLRENPLFGPSSSTLEKLGALEWDKIVHKNEAWRLITCIWLHAGVIHLLANMLSLV
At07250   LSFEPLRTNPLFGPSSHTLEKLGALEWSKVVEKKEGWRLLTCIWLHAGVIHLGANMLSLV
Ote14     FSFQPLGENPMFGPSSSTLLRFGGLNWDKVVHHHQGWRLISCIWLHAGLIHLVVNMLCLV
At63120   FSFQPLKENPLFGPSSSTLEKMGALEWRKVVHEHQGWRLLSCMWLHAGIIHLLTNMLSLI
Os35730   FSFQPLRENPLLGPSSATLQKMGALDWNKVVHQHQGWRLISCIWLHAGLIHLVVNMLSLL
Os43320   FAFQPLRENPLLGPSSATLEKMGALDWAKVVHQHQAWRLISCIWLHAGLIHLIVNMLSLL
Os47840   FAFQPLKENPLLGPSSATLLKMGALDVTKVVHGHQGWRLITCIWLHAGVVHLLINMLCLL
          ::*.*:  **::**** :* ::*.*:  ::*. ::.***::*:***.*:.**  ***.*:
```

Fig. 4.15 A portion of the multiple sequence alignment of the rhomboid sequences from the three plants with the insert region indicated in yellow. The first and the second TM-helices have been indicated in green

4.3.6.3 Class III—I Rhom-like Class

The GXSX motif is replaced with GPYX which is characteristic of the i-Rhom class found in animals. The WR motif is replaced with a WK motif, and the H is replaced with an I. The sequences in the class do not have the catalytic dyad of S and H and are expected to be inactive. The cluster contains a single sequence from Ote, Ote18.

4.3.6.4 Class IV—Other Inactive Homologs

The GXSX motif is replaced with GFX, WR with WN and H with I. Since the catalytic dyad is not conserved, the sequences are not expected to be active proteases. The sequences from Ote in this class are Ote6, Ote8, Ote16, Ote17, Ote20, Ote21, and Ote22. The orthologs are predicted to be mixed other secretases.

Some of the hits were found to be paralogs like

1. Ote8 and Ote17,
2. Ote2, Ote3, and Ote15,
3. Ote9 and Ote12,
4. Ote7 and Ote11.

They picked other *O. tenuiflorum* hits in the all-against-all BLAST searches and clustered with these hits in the phylogeny.

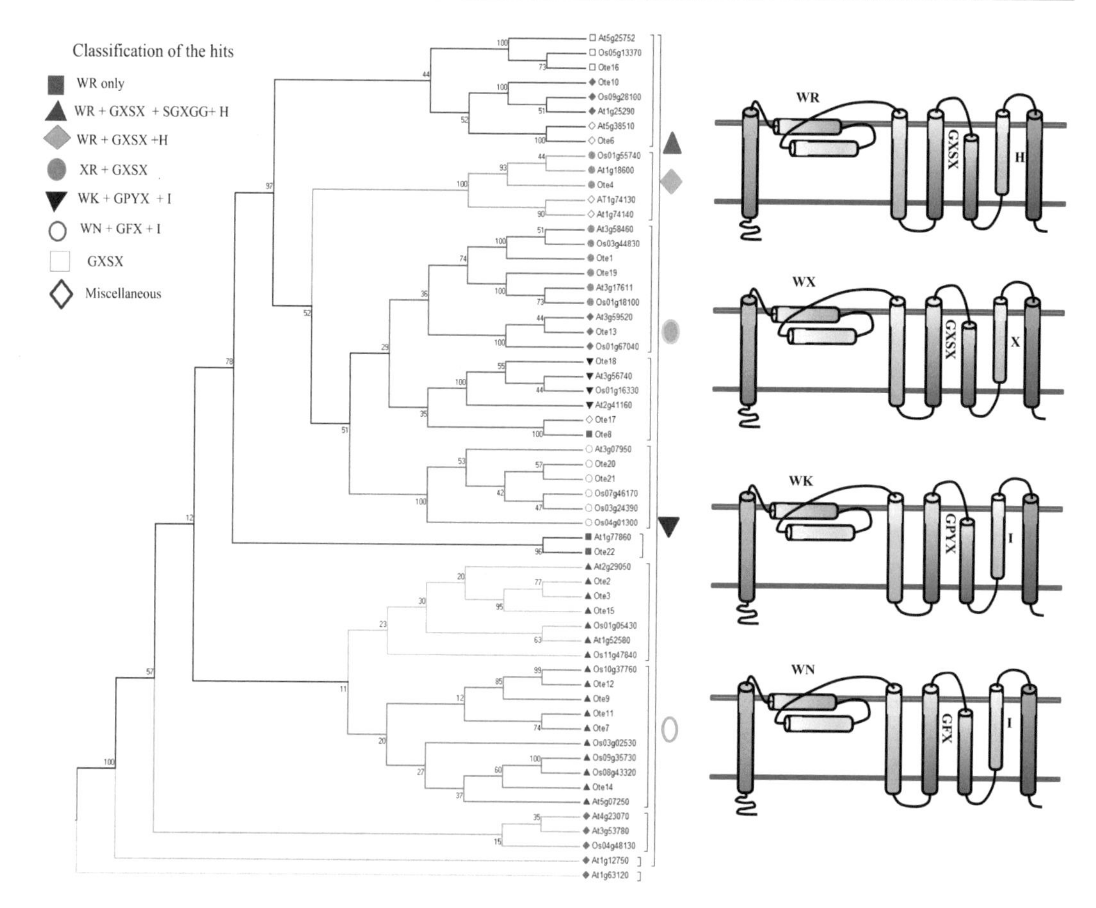

Fig. 4.16 Phylogenetic tree showing the relationship between *O. tenuiflorum* rhomboids and *A. thaliana* and *O. sativa* orthologs. The tree was constructed using NJ from Phylip package. The clusters have distinct functionally important residues which have been indicated in the same color code in the right. Class I members cluster together (as shown in dark blue filled triangle and green filled diamond symbols). Class II members cluster together (marked in orange ellipse symbol), while Class III members are indicated in inverted-brown-triangle symbol. Clusters belonging to class IV members are shown in open-green-ellipse symbol

4.4 Conclusion

The knowledge of omics data would provide clues to understand the structural and functional diversity of plant enzymes. The genome-wide search for TPS in Ote showed clear clustering with the existing classification. The results also showed that there is incomplete structural data for TPS especially in the subfamily B, D, E/F, and H. Hence, these Ote entries can be used to set potential target for crystallographic experiments, which will be helpful to identify more sequences of this type of subfamily. Among the interesting enzymes belonging to TPS, squalene synthase has shown lots of distinctive features. It shows the signature folding pattern, and it is known to follow the typical TPS reaction mechanism. However, it does not show characteristic motif DDXXD or DXDD, which is seen conserved throughout the entire superfamily. This was further confirmed in phylogenetic analysis that squalene synthase sequences cluster separately, and it will be interesting to study their evolutionary links.

In superfamily-based alignment, apart from residue conservation at superfamily level, family-specific conserved residues were also observed. Apart from the strong motifs (discussed in the literature), additional class-specific ones have been noticed in the alignment. All these data can pave for further work in finding the role of individual residue at the catalytic sites.

Among the patterns, an evolutionary flow can be seen in catalytic important residues and gradual change of amino acids from NSE motif to DTE motif from bacterial, fungal to plant-based TPS. This flow is observed in superfamily terpenoid synthase (SCOP ID: 48576).

Both TPS and CYP450 enzyme combination produce a few hundred chemically diverse metabolites that form terpenes. Some of the terpene molecules and the basic scaffold of the chemical structure are biotechnologically important in the form of novel drug molecules, potential nutraceutical products. Studying the sequence variations not only helps in understanding the mechanism of how these enzymes functions, but also provide a tool for further engineering these enzymes to obtain products of commercial importance. The prescribed approach for identifying and classifying the CYP450 families in Ote would help to discover all P450 s in other sequenced plant genomes.

In this study, four distinct classes of rhomboids were identified. The classification correlated with the cellular localization and had unique motifs for the functionally important regions. The first class had all the functionally important motifs conserved—WR in the loop 1, GXSX in the fourth TM-helix and H in the sixth TM-helix. The second class has the motif WX, GXSX and variable residues in the place of H. Few plant rhomboids have the characteristic motif GPYX motif which resembles the i-Rhom motif of GPXX. However, the loop regions characteristic of the i-Rhoms are not present in the plant sequences. The plant sequences also have a WK motif instead of WR and I in place of the catalytic H in i-Rhoms-like proteins. The sequences with i-Rhom-like motifs have not been described in plants earlier. Another class of rhomboids was also found with the motifs WN, GFXX, and I in place of the functionally important residues WR, GXSX, and H. Based on conservation of functionally important residues, it was predicted that the first class will be functionally active serine proteases indicating that only mixed other secretase type rhomboids may be catalytically active in plants. Enzymatically, inactive rhomboids have shown to be involved in processes like transport of the proteins across membranes. While the rhomboids have a low sequence identity, the fact that there is an expansion of these proteins in plants and some rhomboids are expressed in a tissue-specific way indicates that these proteins have important functions in plants.

Overall, the sequence and structure details of plant enzyme superfamilies in Tulsi genome, along with the suggested methodology, would help to perform genome-wide search in more plant genomes in future.

References

Abdallah I, Quax WJ (2017) A glimpse into the biosynthesis of terpenoids. KnE Life Sci 3(5):81–98

Afendi FM, Okada T, Yamazaki M, Hirai-Morita A, Nakamura Y, Nakamura K, Ikeda S, Takahashi H, Altaf-Ul-Amin M, Darusman LK, Saito K, Kanaya S (2012) KNApSAcK family databases: Integrated metabolite-plant species databases for multifaceted plant research. Plant Cell Physiol 53(2):1–12

Altschul SF, Gish W, Miller W, Myers EW, Lipman DJ (1990) Basic local alignment search tool. J Mol Biol 215:403–410

Baerenfaller K, Hirsch-Hoffmann M, Svozil J, Hull R, Russenberger D, Bischof S, Baginsky S (2011) pep2pro: a new tool for comprehensive proteome data analysis to reveal information about organ-specific proteomes in Arabidopsis thaliana. Integrative Biology: Quantitative Biosciences from Nano to Macro 3 (3), 225–237 https://doi.org/10.1039/c0ib00078g

Baldi P, Chauvint Y, Hunkapiller T, Mcclureii M (1994) Hidden Markov models of biological primary sequence information (multiple sequence alignments/protein modeling/adaptive algorithms/sequence Classification). Biochemistry 91:1059–1063

Biegert A, Soding J (2009) Sequence context-specific profiles for homology searching. Proc Natl Acad Sci USA 106(10):3770–3775

Bohlmann J, Meyer-Gauen G, Croteau R (1998) Plant terpenoid synthases: molecular biology and phylogenetic analysis. Proc Natl Acad Sci USA 95(8):4126–4133

Boutanaev AM, Moses T, Zi J, Nelson DR, Mugford ST, Peters RJ, Osbourn A (2015) Investigation of terpene diversification across multiple sequenced plant genomes. Proc Natl Acad Sci USA 112(1):E81–E88

Camacho C, Coulouris G, Avagyan V, Ma N, Papadopoulos J, Bealer K, Madden TL (2009) BLAST+: architecture and applications. BMC Bioinform 9:1–9
Caspi R, Altman T, Dreher K, Fulcher CA, Subhraveti P (2016) The MetaCyc database of metabolic pathways and enzymes and the BioCyc collection of pathway/genome databases. Nucl Acids Res 44(D1): D471–D480
Chapple C (1998) Molecular-genetic analysis of plant cytochrome P450-dependent monooxygenases. Annu Rev Plant Physiol Plant Mol Biol 49:311–343
Chen F, Tholl D, Bohlmann J, Pichersky E (2011) The family of terpene synthases in plants: a mid-size family of genes for specialized metabolism that is highly diversified throughout the kingdom. Plant J 66 (1):212–229
Degenhardt J, Köllner TG, Gershenzon J (2009) Monoterpene and sesquiterpene synthases and the origin of terpene skeletal diversity in plants. Phytochemistry 70 (15–16):1621–1637
Eddy SR (1998) Profile hidden Markov models. Bioinformatics 14(9):755–763
Farnsworth NR (1988) Screening plants for new medicines. In: Wilson EO, Peter FM (eds) Biodiversity. National Academies Press, Washington, D.C.
Felsenstein J (1981) Evolutionary trees from DNA sequences: a maximum likelihood approach. J Mol Evol 17(6):368–376
Fox NK, Brenner SE, Chandonia JM (2014) SCOPe: Structural Classification of Proteins—extended, integrating SCOP and ASTRAL data and classification of new structures. Nucleic Acids Res 42(D1):D304–D309. https://doi.org/10.1093/nar/gkt1240
García-Lorenzo M, Sjödin A, Jansson S, Funk C (2006) Protease gene families in *Populus* and *Arabidopsis*. BMC Plant Biol 6:1–24
Gertz EM, Yu Y-K, Agarwala R, Schäffer AA, Altschul SF (2006) Composition-based statistics and translated nucleotide searches: improving the TBLASTN module of BLAST. BMC Biol 4:41
Gotoh O (1992) Substrate recognition sites in cytochrome P450 family 2 (CYP2) proteins inferred from comparative analyses of amino acid and coding nucleotide sequences. J Biol Chem 267(1):83–90
Gotoh O (2012) Evolution of cytochrome P450 genes from the viewpoint of genome informatics. Biol Pharm Bull 812(356):812–817
Hofmann K (1993) TMBASE-A database of membrane spanning protein segments. Biol Chem Hoppe-Seyler 374, 166. Retrieved from https://ci.nii.ac.jp/naid/10007774832/en/
Hruz T, Laule O, Szabo G, Wessendorp F, Bleuler S, Oertle L, Zimmermann P (2008) Genevestigator V3: A Reference Expression Database for the Meta-Analysis of Transcriptomes. Adv Bioinform. https://doi.org/10.1155/2008/420747
Jones DT (1999) Protein secondary structure prediction based on position-specific scoring matrices. J Mol Biol 292:195–202
Käll L, Krogh A, Sonnhammer EL (2004) A combined transmembrane topology and signal peptide prediction method. J Mol Biol 338(5):1027–1036
Kampranis SC, Ioannidis D, Purvis A, Mahrez W, Ninga E, Katerelos NA, Anssour S, Dunwell JM, Degenhardt J, Makris AM, Goodenough PW, Johnson CB (2007) Rational conversion of substrate and product specificity in a *Salvia* monoterpene synthase: Structural insights into the evolution of terpene synthase function. Plant Cell 19(6):1994–2005
Kanaoka MM, Urban S, Freeman M, Okada K (2005) An Arabidopsis Rhomboid homolog is an intramembrane protease in plants. FEBS Letters. https://doi.org/10.1016/j.febslet.2005.09.049
Kmiec-Wisniewska B, Krumpe K, Urantowka A, Sakamoto W, Pratje E, Janska H (2008) Plant mitochondrial rhomboid, AtRBL12, has different substrate specificity from its yeast counterpart. Plant Mol Biol 68(1–2):159–171
Knopf RR, Adam Z (2012) Rhomboid proteases in plants - still in square one? Physiol Plant 145(1):41–51
Kong DX, Guo MY, Xiao ZH, Chen LL, Zhang HY (2011) Historical variation of structural novelty in a natural product library. Chem Biodivers 8(11):1968–1977
Koonin EV, Makarova KS, Rogozin IB, Davidovic L, Letellier MC, Pellegrini L (2003) The rhomboids: a nearly ubiquitous family of intramembrane serine proteases that probably evolved by multiple ancient horizontal gene transfers. Genome Biol 4(3):R19
Lemberg MK, Freeman M (2007) Functional and evolutionary implications of enhanced genomic analysis of rhomboid intramembrane proteases. Genome Res 17 (11):1634–1646
Lemberg MK, Menendez J, Misik A, Garcia M, Koth CM, Freeman M (2005) Mechanism of intramembrane proteolysis investigated with purified rhomboid proteases. EMBO J 24(3):464–472
Li Q, Zhang N, Zhang L, Ma H (2015) Differential evolution of members of the rhomboid gene family with conservative and divergent patterns. New Phytol 206(1):368–380
Marchler-Bauer A, Panchenko AR, Shoemaker BA, Thiessen PA, Geer LY, Bryant SH (2002) CDD: a database of conserved domain alignments with links to domain three-dimensional structure. Nucl Acids Res 30(1):281–283
Martin VJ, Pitera DJ, Withers ST, Newman JD, Keasling JD (2003) Engineering a mevalonate pathway in *Escherichia coli* for production of terpenoids. Nat Biotechnol 21:796–802
Mayer U, Nüsslein-Volhard C (1988) A group of genes required for pattern formation in the ventral ectoderm of the Drosophila embryo. Genes Dev 2(11):1496–1511
Nebert DW, Nelson DR, Coon MJ, Estabrook RW, Feyereisen R, Fujii-Kuriyama Y, Gonzalez FJ, Guengerich FP, Gunsalus IC, Johnson EF et al (1991) The P450 superfamily: update on new sequences, gene mapping, and recommended nomenclature. DNA Cell Biol 10(1):1–14

Nelson DR (2009) The cytochrome p450 homepage. Hum Genom 4(1):59–65
Pazouki L, Niinemets Ü (2016) Multi-substrate terpene synthases: their occurrence and physiological significance. Front Plant Sci 7:1019
Radivojac P, Clark WT, Oron TR, Schnoes AM, Wittkop T et al (2013) A large-scale evaluation of computational protein function prediction. Nat Meth 10(3):221–227
Rambaut A (2009). FigTree. Tree Figure Drawing Tool. http://Tree.Bio.Ed.Ac.Uk/Software/Figtree/
Schuler MA (1996) The role of cytochrome P450 monooxygenases in plant-insect interactions. Plant Physiol 112(4):1411–1419
Sonnhammer EL, Eddy SR, Birney E, Bateman A, Durbin R (1998) Pfam: multiple sequence alignments and HMM-profiles of protein domains. Nucl Acids Res 26(1):320–322
Stevenson LG, Strisovsky K, Clemmer KM, Bhatt S, Freeman M, Rather PN (2007) Rhomboid protease AarA mediates quorum-sensing in *Providencia stuartii* by activating TatA of the twin-arginine translocase. Proc Natl Acad Sci 104(3):1003–1008
Tholl D (2006) Terpene synthases and the regulation, diversity and biological roles of terpene metabolism. Curr Opin Plant Biol 9(3):297–304
Thompson EP, Llewellyn Smith SG, Glover BJ (2012) An arabidopsis rhomboid protease has roles in the chloroplast and in flower development. J Exp Bot 63 (10):3559–3570
Tripathi LP, Sowdhamini R (2006) Cross genome comparisons of serine proteases in arabidopsis and rice. BMC Genom 7:1–31
Upadhyay AK, Chacko AR, Gandhimathi A, Ghosh P, Harini K, Joseph AP, Joshi AG (2015) Genome sequencing of herb Tulsi (*Ocimum tenuiflorum*) unravels key genes behind its strong medicinal properties. BMC Plant Biol 15(1):1–20
Urban S, Freeman M (2003) Substrate specificity of rhomboid intramembrane proteases is governed by helix-breaking residues in the substrate transmembrane domain. Mol Cell 11(6):1425–1434
Urban S, Lee JR, Freeman M (2001) Drosophila rhomboid-1 defines a family of putative intramembrane serine oroteases. Cell 107(2):173–182
Wasserman JD, Urban S, Freeman M (2000) A family of rhomboid-like genes : Drosophila Rhomboid-1 and Roughoid/ Rhomboid-3 cooperate to activate EGF receptor signaling. Genes Dev 14(13):1651–1663
Zhang Z, Schäffer AA, Miller W, Madden TL, Lipman DJ, Koonin EV, Altschul SF (1998) Protein sequence similarity searches using patterns as seeds. Nucl Acids Res 26(17):3986–3990

Systematic Position, Phylogeny, and Taxonomic Revision of Indian *Ocimum*

5

Amit Kumar, Ashutosh K. Shukla, Ajit Kumar Shasany and Velusamy Sundaresan

Abstract

The genus *Ocimum* belongs to the family Lamiaceae, and most popular herb/shrub known for its immense value belongs to their medicinal properties and aroma profile. The presence of vast morphological as well as chemical variability within the genus, its taxonomy, and phylogenetic relationships is still debatable. Adding to this, chemotypes within the species, which are more or less similar in morphology, increases the complexity to higher level. Nowadays, the use of molecular markers, viz. PCR and sequence-based markers, became an essential tool to assess interspecific relationships as they are highly efficient and provide more variable and informative insight into evolutionary rates. The present article presents the detailed taxonomic description as well as phylogenetic relationships among the Indian *Ocimum* species.

A. Kumar · V. Sundaresan (✉)
Plant Biology and Systematics, CSIR—Central Institute of Medicinal and Aromatic Plants, Research Centre, Bengaluru 560065, India
e-mail: vsundaresan@cimap.res.in

A. K. Shukla · A. K. Shasany
Biotechnology Division, CSIR—Central Institute of Medicinal and Aromatic Plants, Lucknow 226015, India

5.1 Introduction

Ocimum L. (1753: 597, 1754: 259), Loureiro (1790: 369), as "*Ocymum*", Bentham (1830: 13, as "*Ocymum*", 1832: 1, as "*Ocymum*", 1848: 31, 1876: 1171), Hooker f. (1885: 607), Briquet (1897: 369), Ridley (1923: 643), Kudo (1929: 112), Doan (1936: 918), Mukerjee (1940: 17), Bakhuizen van den Brink f. (1965: 638), Keng (1969: 125, 1978: 376), Cramer (1981: 111), Paton (1992: 409), Paton et al. (1999: 26), Suddee et al. (2005:24).

The genus *Ocimum* L., collectively called as "Basil" considered as one of the largest genera of the family Lamiaceae (sixth largest family), includes an important economic and medicinal group of herbs or undershrubs, widely distributed in the tropical, subtropical, and warm temperate regions of the world (Paton et al. 1999). The genus *Ocimum* has long been acclaimed for its diversity as a source of essential oils, its flavor and delicacy as a spice, and its beauty and fragrance as an ornamental (Simon et al. 1990). The species of *Ocimum* is characterized by a great chemical variability consisting of monoterpenes, sesquiterpenes, and phenylpropanoids affecting the commercial value of this genus. The oil components of *Ocimum* have been found to be produced by two different biochemical pathways, viz. shikimic acid pathway (phenylpropanoids) and mevalonic acid pathway (terpenes). Monoterpenoids (linalool, 1, 8-cineole, camphor,

A. K. Shasany and C. Kole (eds.), *The Ocimum Genome*, Compendium of Plant Genomes, https://doi.org/10.1007/978-3-319-97430-9_5

limonene, citral, geranial, neral) and phenylpropanoids (eugenol, methyl eugenol, methyl chavicol, methyl cinnamate) with noticeable amounts of sesquiterpenoids (mainly β-caryophyllene, β-elemene, β-selinene, α-selinene, β-bisabolene, germacrene D) are reported as prevalent components distributed in different *Ocimum* spp. and their chemotypes (Balyan and Pushpangadan 1988; Vina and Murillo 2003; Verma et al. 2016).

The genus *Ocimum* L. is characterized as usually woody and aromatic, an erect, annual, or perennial herbs or undershrubs. *Stems*—quadrangular. *Leaves*—petiolate, opposite, margin entire or serrate. *Inflorescence*—terminal, simple, branched at base, clearly interrupted verticils, cymes sessile, and unbranched. *Flowers*—small (whorls of 6–10 on racemes), bracteolate, bracts small, caducous, clawed, or subsessile. *Calyx*—bilabiate, ovoid or campanulate, posterior lip broad, anterior lip four-lobed (equal or longer than posterior). *Corolla*—bilabiate, tube short, declinate, entire; posterior lip erect, subequally four-lobed; anterior lip longer. *Stamens*—four didynamous, declinate, exserted, subequal or with the anterior ones slightly longer, posterior attached near the base of corolla tube, filaments free, anthers synthecous. *Disk*—entire or three–four-lobed. Ovary—glabrous, style declinate, bifid with branches usually subequal, subulate and flattened. *Nutlets*—obovoid, oblong, ellipsoid or subglobose, often mucilaginous when wet (Suddee et al. 2005).

Occurrence of polyploidy and possibility of inter- and intraspecific hybridization within the genus *Ocimum* lead to a large number of subspecies, varieties, and forms, which differ in essential oil composition and morphological characters. Estimates of number of species within the genus *Ocimum* vary from 30 (Paton 1992) to 160 (Pushpangadan and Bradu 1995). Balyan and Pushpangadan (1988) reported nine species of *Ocimum* from India, of which three species were described as exotic. However, further revision of the genus *Ocimum* carried out by Suddee et al. (2005) merged *O. canum* with *O. americanum*. Moreover, study carried out by Carovic-Stanko et al. (2010) reported that the separate species rank of *O. minimum* was not justified and suggested a subspecies or even a variety of *O. basilicum*. Hence, seven species of *Ocimum* are currently recognized from India which include *O. tenuiflorum* L., *O. basilicum* L., *O. gratissimum* L., *O. kilimandscharicum* Gurke, *O. americanum* L., *O. filamentosum* Forssk., and *O. africanum* Lour.

5.2 Key to *Ocimum* Species (Source: Suddee et al. 2005)

1. Teeth of lateral fruiting calyx lobes distinct, deltoid, or lanceolate; bract without bowl-like gland developing.

    2. Throat of fruiting calyx closed by the upcurved two median teeth of anterior lip, anterior much shorter than posterior; mostly shrubs or undershrubs; nutlets large, not mucilaginous when wetted.
    O. gratissimum
    2. Throat of fruiting calyx open, the two median teeth of anterior lip as long as or longer than posterior; mostly herbs with stems woody at base.

        3. Calyx tube glabrous or thinly covered with minute glandular hairs inside; nutlets small, nearly smooth, not mucilaginous when wetted.
        O. tenuiflorum
        3. Calyx tube with a ring of hairs at throat inside; nutlets producing mucilage when wetted.

            4. Flower with pedicel nearly as long as calyx; appendage of posterior stamens hairy; nutlets large.
            O. kilimandscharicum
            4. Flower with pedicel much shorter than calyx; appendage of posterior stamens glabrous.

                5. Fruiting calyx up to 5 mm long; corolla 4–5.5 mm long.

                    6. Fruiting calyx 2–3 mm long; stem internodes with short

adpressed or retrorse hairs; nutlets small.
O. americanum

6. Fruiting calyx 4–5.5 mm long; stem internodes with long, spreading and sometimes retrorse hairs; nutlets small, tuberculate.
O. africanum

5. Fruiting calyx 6–8 mm long; corolla 7–8 mm long; nutlets large.
O. basilicum

1. Teeth of lateral fruiting calyx lobes obscure, fringed with a row of many minute teeth; bract caducous, with bowl-like gland developing in the scar; nutlets orbicular, mucilaginous when wetted.
O. filamentosum

5.2.1 *Ocimum tenuiflorum*

L. (1753: 597); Bentham (1832: 12; 1848: 39), Keng (1978: 378), Press (1982: 160), Keng (1990: 197), Paton (1992: 432), Ho (1993: 1067), Phuong (1995: 41), Budantsev (1999: 29), Clement (1999: 1002), Phuong (2000: 78), Suddee et al. (2005:26).

Ocimum tenuiflorum L. (syn. *Ocimum sanctum* L.) is an important sacred medicinal plant that grows wild and native to India, commonly known as "holy basil" and "Tulsi", worshiped for over more than 3000 years (Fig. 5.1a). This species is considered as "The Queen of Herbs," "The Incomparable One," and "The Mother Medicine of Nature" in traditional herbal medicine as well as mentioned in Ayurvedic text of Charaka Samhita (Rastogi et al. 2015). Two types of *O. tenuiflorum* are commonly found in nature: (i) Tulsi plants with green leaves known as Ram Tulsi and (ii) Tulsi plants with purple leaves known as Krishna Tulsi and is used as medicinal plants in day-to-day practice in Indian homes for various ailments. *O. tenuiflorum* with a basic chromosome number of $x = 9$ has been reported to possess antifertility, anticancer, antidiabetic, antifungal, antimicrobial, cardioprotective, analgesic, antispasmodic, and adaptogenic activity. The essential oil of leaf contains terpenes and phenylpropenes (eugenol, euginal, urosolic acid, carvacrol, limatrol, caryophyllene, methyl carvicol), which are synthesized and deposited in glandular peltate trichomes (Rastogi et al. 2015).

The characteristic features of *O. tenuiflorum* are described as short-lived perennial herbs up to 1 m tall. *Stems*—round–quadrangular and hirsute. *Leaves*—elliptic, oblong, or ovate–oblong, 5–45 × 5–20 mm, apex obtuse or acute, base obtuse, margin coarsely serrate, glandular-punctate, dorsally glabrous. *Petiole*—4–15 mm long, hirsute. *Inflorescence*—lax or dense, verticils 5–10 mm apart, axis hirsute, bracts ovate 2–3 × 2–4 mm, pedicels 2.5–4 mm long in fruit. *Calyx*—1–1.5 mm long at anthesis, 3–4 mm long in fruit, posterior lip rounded, anterior lip with two median lanceolate teeth curved upwards, throat open, tube with patent hairs. *Corolla*—2–3 mm long purplish red, lobes pubescent on back, posterior lip with the two oblong lateral lobes, anterior lip obovate–oblong, tube glabrous on both sides. *Stamens*—with posterior pair hirsute at base. *Nutlets*—brown, ovoid–oblong, smooth, non-mucilaginous (Suddee et al. 2005).

5.2.2 *Ocimum basilicum*

L. (1753: 597); Lour. (1790: 370), as "*Ocymum*"; sensu auctt. Benth. (1830: 13), *excl. var. pilosum*; Benth. (1832: 4), *excl. var. pilosum* and *var. anisatum*; Benth. (1848: 32), *excl. var. pilosum* and *var. anisatum*; Hook. f. (1885: 608), *excl. syn. O. pilosum* and *O. menthaefolium* (the Indian plant); Muschler and Hosseus (1910: 501), Ridley (1923: 643), Kudo (1929: 113), Merr. (1935: 343), Doan (1936: 919, f 96, 1–3) pro parte, quoad Evrard 2442, Poilane 9773 and 40624, Mukerjee (1940: 18), *excl. syn. O. hispidum* and *O. pilosum*, Keng (1969: 127) pro parte, Murata (1971: 509), Li (1977: 561), *excl. var pilosum*; Keng (1978: 377), pro parte, Cramer (1981: 115),

Fig. 5.1 Genus *Ocimum* and its species

excl. syn. O. americanum L., Phuong (1982: 146), *excl. var pilosum*; Press (1982: 160), Keng (1990: 197), Paton (1992: 423), *excl. syn. O. citriodorum*; Li and Hedge (1994: 296), *excl. var pilosum*, Phuong (1995: 40), Clement (1999: 1001), Paton et at. (1999: 25), Phuong (2000: 88), *excl. var. pilosum*; Suddee et al. (2005:30).

Ocimum basilicum L. popularly known as "Sweet basil" (Fig. 5.1b) with basic chromosome number $x = 12$ has been used as a popular culinary herb and a rich source of essential oils primarily consisting of monoterpenes and phenylpropanoids, which are used to flavor foods, in oral products, and in perfumeries. The major constituents were found to be linalool, methyl chavicol, citral, 1, 8-cineole, camphor, thymol, methyl cinnamate, eugenol, methyl eugenol, methyl isoeugenol, and elemicine. The essential oil has antimicrobial, antifungal, and insect-repelling, anticonvulsant, hypnotic, and antioxidant activities. Medicinally, the plant is used in the treatment of headaches, coughs, diarrhea, constipation, warts, worms, and kidney malfunctions.

Annual or short-lived perennial herbs up to 1 m height. *Stems*—quadrangular, rounded or round–quadrangular, glabrous or sparsely pubescent. *Leaves*—dark green, ovate or elliptic ovate, 15–50 × 5–25 mm, apex acute, base cuneate, margin entire or sparsely serrate, glandular-punctate, glabrous. *Petiole*—20 mm long, pubescent. *Inflorescence*—lax or dense, verticils up to 12 mm apart, axis pubescent. Bracts—ovate, elliptic, elliptic ovate/lanceolate, 6–10 × 2–5 mm, apex acuminate, base cuneate or attenuate, glandular-punctate, pubescent. *Pedicels*—1–2 mm long, pubescent. *Calyx*—campanulate, 4–5 mm long at anthesis, 6–8 mm long in fruit, posterior lip rounded, margin curved, apiculate at apex, anterior lip with two median lanceolate, acuminate teeth, slightly longer than posterior, throat open, tube with sessile glands. *Corolla*—white, purple or white with purple margin, 7–8 mm long, lobes obscurely crenate, pubescent, posterior lip with two median oblong lobes and two lateral broadly oblong lobes, anterior lip boat-shaped, tube straight, glabrous. *Stamens*—with posterior having a transverse process of tufted hairs near base. *Nutlets*—dark brown, oblong or ovoid–ellipsoid, 2–2.5 × 1–1.5 mm, highly mucilaginous (Suddee et al. 2005).

5.2.3 *Ocimum* kilimandscharicum

Baker ex Gurke (1895: 349), Paton (1992: 422), Paton et al. (1999: 24), Suddee et al. (2005:27).

Ocimum kilimandscharicum (Fig. 5.1c) is an economically important medicinal perennial herb and is widely used for the treatment of various ailments including colds, coughs, abdominal pains, measles, and diarrhea (Obeng-Ofori et al. 1998). It is also considered as a source of essential oils containing biologically active constituents that act as insect repellents, particularly against mosquitoes and storage pests (Kokwaro 1976) or show antibacterial (Prasad et al. 1986) and antioxidant activity (Hakkim et al. 2008).

Perennial aromatic undershrub up to 1 m tall. *Stems*—four-angled, woody at base, pubescent. *Leaves*—simple, opposite-decussate, elliptic ovate to oblong, 35–60 × 15–30 mm; petiole hirsute, 10–28 mm; leaf base cuneate, unequal; margin serrate, apex acute, pubescent. *Flowers*—small, dense, verticillate, in terminal hairy racemes; verticils 3–15 mm apart; bracts sessile, ovate to lanceolate, hirsute, 2–3 mm; spike 14–30 cm, pedicel 2–3 mm. *Calyx*—bilipped, campanulate, hirsute, posterior lip large, broad, rounded, decurrent on tube, apex mucronate; anterior lip four-toothed, with two median lanceolate teeth slightly move upwards, slightly longer than two lateral teeth, more or less equal to posterior lip; throat open, tube pubescent with ring of hair at throat base. *Corolla*—campanulate, purplish white, two-lipped, 7–9.5 mm long including corolla tube, lobes pubescent dorsally; upper lip truncate, four fid, with two ovate–oblong smaller median lobes apparently joined together and slightly longer than two lateral broader lobes; lower lip elliptic-oblong, comparatively white. *Stamens*—four, declinate, in two pairs, exserted, posterior pair with a fascicle of hair near the base, anther cells confluent, filaments free. *Style*—thinly bifid. *Nutlets*—ovoid–oblong, black; mucilaginous (Suddee et al. 2005).

5.2.4 *Ocimum americanum*

L. (1755: 15), as "*Ocymum*"; Backer and Bakhuizen van den Brink f. (1965: 640), Keng (1969: 126), Li (1977: 560), Keng (1978: 376), Press (1982: 160), Keng (1990: 197), Li and Hedge (1994: 296), Phuong (1995: 40), Paton and Putievsky (1996: 513) pro parte; Clement (1999: 1001), Phuong (2000: 87), Suddee et al. (2005:28).

Ocimum americanum L. (syn. *O. canum* Sims), commonly known as "Hoary basil," occurs as weed throughout tropical and subtropical parts of India (Fig. 5.1d). It is not often used as a culinary herb, but more often as a medicinal plant. The essential oils found in this species have strong fungicidal (Singh and Dwivedi 1987) and insecticidal (Weaver et al. 1991) activity. Six chemotypes of *O. americanum* containing methyl cinnamate, camphor, methyl chavicol, linalool, eugenol, farnesol, and citral have been reported from different parts of the world.

Annual or short-lived perennial herbs up to 10–40 cm tall. *Stems*—round–quadrangular with retrorse hairs. *Leaves*—lanceolate or ovate–lanceolate, 5–25 × 5–15 mm, apex acute, base cuneate, margin entire or sparsely serrate, glandular-punctate, glabrous. *Petiole*—2–15 mm long, pubescent. *Inflorescence*—lax, verticils up to 10 mm apart, axis densely pubescent, bracts ovate, 3–4 mm long, apex acute or acuminate, base attenuate, glandular-punctate, pedicels 1–2 mm long, recurved, pubescent. *Calyx*—campanulate, 1.5–2 mm long, posterior lip rounded, anterior lip with two median lanceolate, acuminate teeth, slightly longer than posterior, lateral teeth broad deltoid, acute, almost equal to posterior, throat open, tube with or without sessile glands. *Corolla*—white or light purple, 4–5 mm long, lobes entire, pubescent, posterior lip with two median oblong lobes and two lateral obovate–oblong lobes, anterior lip boat-shaped, tube straight, glabrous. *Stamens*—with posterior having a glabrous transverse process near base. *Nutlets*—black, oblong, 0.8–1 mm long, minutely tuberculate, mucilaginous (Suddee et al. 2005).

5.2.5 *Ocimum africanum*

Lour. (1790: 370), as "*Ocymum*", Merrill (1935: 343), Suddee et al. (2005:28).

Ocimum africanum Lour. (syn. *O.* × *citriodorum*) the Lemon basil (Fig. 5.1e) is known to be a hybrid of *O. basilicum* L. and *O. americanum* L. (Paton and Putievsky 1996). The herb is known for its strong fragrant lemon scent, making it a useful cooking ingredient in Arabic, Indonesian, Lao, Persian, and Thai cuisine. Grayer et al. (1996) found citral (= geranial + neral) to be a major constituent in *O. africanum*. Essential oil of *O. africanum* was reported to be highest antibacterial activity among the other species of *Ocimum* (Carovic-Stanko et al. 2010). Essential oil of *O. africanum* was also reported to possess anticholinesterase Activity (Tadros et al. 2014).

Annual or short-lived perennial herbs up to 10–50 cm tall. *Stems*—round–quadrangular, densely pubescent. *Leaves*—elliptic, lanceolate, ovate–lanceolate or ovate–oblong, 5–35 × 5–20 mm, apex acute, base cuneate or obtuse, margin entire or sparsely serrate, glandular-punctate, glabrous. *Petiole*—2–20 mm long, pubescent. *Inflorescence*—lax, verticils up to 10 mm apart. Bracts—ovate, 5 mm long, apex acute or acuminate, base attenuate, margin pilose, glandular-punctate. Pedicels—1–2.5 mm long, recurved, pubescent. *Calyx*—campanulate, 1.5–2.5 mm long, posterior lip rounded, anterior lip with two median lanceolate, acuminate teeth, longer than posterior, lateral teeth broad deltoid, acute, almost equal to posterior, throat open, tube with or without sessile glands. *Corolla*—white or light purple, 4–5.5 mm long, lobes entire, pubescent, posterior lip with two median oblong lobes and two lateral obovate–oblong lobes, anterior lip boat-shaped, tube straight, glabrous.

Stamens—with posterior having a glabrous transverse process near base. *Nutlets*—black, narrowly oblong, 1–1.5 mm long, minutely tuberculate, mucilaginous (Suddee et al. 2005).

5.2.6 *Ocimum gratissimum*

L. (1753: 1197), as "*Ocymum*"; Jacquin (1792: 7, t. 495.), as "*Ocymum*", Bentham (1830: 14, as "*Ocymum*", 1832: 7, as "*Ocymum*", 1848: 34); Hooker f. (1885: 608), Ridley (1923: 644), Doan (1936: 919, f. 96, 4–5), Mukerjee (1940: 20), Keng (1969: 128), Li (1977: 564), Keng (1978: 377), Cramer (1981: 112), Phuong (1982: 146), Press (1982: 160), Paton (1992: 411), Ho (1993: 1067), Phuong (1995: 41), Paton et al. (1999: 26), Phuong (2000: 91), Suddee et al. (2005: 25).

Ocimum gratissimum L. commonly known as "tree basil" (Fig. 5.1f), native to West Africa (Pino et al. 1996) has been used extensively in the traditional system of medicine in many countries. Decoction of the leaves or the whole herb is used in the treatment of fever, as a diaphoretic, as a laxative, and as an anthelmintic medicine (Charles and Simon 1992). *O. gratissimum* has basic chromosome number $x = 10$ and is subdivided into two varieties: *O. gratissimum* var. *gratissimum* L., which has a hairy stem, pubescent leaves, inflorescence lax or dense; and *O. gratissimum* var. *macrophyllum* Briq., with glabrous stem and leaves, inflorescence lax (Paton 1992). *O. viridie* ($2n = 40$) and *O. suave* ($2n = 48$) were considered as synonyms of *O. gratissimum* var. *gratissimum* (Paton 1992). Based on the essential oil composition, six chemotypes of *O. gratissimum* have been reported: thymol type, eugenol type, citral type, ethyl cinnamate type, linalool type (Pino et al. 1996), and geraniol type (Charles and Simon 1992). Essential oil of *O. gratissimum* showed antagonist activity against a number of Gram-positive, Gram-negative bacteria as well as against protozoan (Tokin 1944).

Perennial undershrubs or shrubs up to 2 m tall. *Stems*—round–quadrangular, woody at base, glabrous. *Leaves*—ovate or elliptic ovate, 25–100 × 12–60 mm, apex acute, base cuneate or attenuate, margin sparsely serrate, glandular-punctate, glabrous. *Petiole*—10–50 mm long, glabrous or pubescent. *Inflorescence*—lax or dense, verticils up to 10 mm apart, axis glabrous, bracts caducous, ovate with very broad base, 2–5 mm long, apex acuminate, base cuneate, margin ciliate, pubescent, three nerves at base. *Pedicels*—2–3 mm long, recurved, pubescent. *Calyx*—2–3 mm long at anthesis, 3–4 mm long in fruit, posterior lip rounded, apex pointed, anterior lip shorter than posterior, two lateral teeth level with or below two median teeth, throat closed, tube with or without sessile glands. *Corolla*—greenish white, 3–4 mm long, lobes equal, obsecurely crenate, posterior lip oblong, anterior lip boat-shaped, tube straight, pubescent outside, glabrous inside. *Stamens*—with posterior having a transverse process near base. *Nutlets*—brown, subglobose, minutely tuberculate, mucilaginous (Suddee et al. 2005).

5.2.7 *Ocimum filamentosum*

Forsskål (1775: 108), as "*Ocymum*", sensu auctt. Bentham (1832: 9, 1848: 36), *excl. syn.*; Paton et al. (1999: 30), Suddee et al. (2005:31).

Ocimum filamentosum Forssk. (syn. *O. adscendens* Willd.) is a lesser known species of genus *Ocimum* (Fig. 5.1g). Major constituents of the *O. filamentosum* essential oil were reported to be eugenol (47.6%), (E)-caryophyllene (15.7%), β-elemene (11.3%), caryophyllene oxide (2.8%), germacrene D (1.7%), β-copaene (1.6%), α-humulene (1.2%), borneol (1.0%), and α-selinene (1.0%) (Verma et al. 2016).

Perennial herbs more than 30 cm tall. *Stems*—simple, round–quadrangular. *Leaves*—elliptic ovate, or obovate–lanceolate, 20–45 × 7–15 mm, apex obtuse or acute, base cuneate or attenuate, margin entire or obscurely serrate, sessile glands, dorsally glabrous. *Petiole*—5 mm long, pubescent. *Inflorescence*—lax, verticils up to 16 mm apart, axis hirsute, bracts sessile, lanceolate, 5–6 × 1.5–2.5 mm, apex acute, margin entire, pubescent, pedicels 1 mm long during anthesis, 2–2.5 mm in fruit, pubescent. *Calyx*—campanulate, 4.5 mm long, posterior lip rounded, anterior lip with two median bristle like teeth, lateral teeth absent, throat open, tube with patent hairs outside, glabrous inside. *Corolla*—10–12 mm long pinkish white, lobes obtuse, pubescent on back, posterior lip four-lobed, anterior lip entire, slightly concave, tube pubescent. *Stamens*—twice as long as corolla. *Nutlets*—light brown or yellow, 1.2–1.5 mm diameter, ovoid–orbicular, smooth, mucilaginous when wet.

5.3 Phylogenetic Relationships

In context to the phylogenetic relationship, the genus *Ocimum* is very complicated due to the presence of huge morphological variability, intra- and interspecific genetic diversity along with vast chemical variations. According to previous reports on classification of *Ocimum*, the genus is placed in Tribe *Ocimeae* Benth. subtribe *Ociminae* which is characterized by a small flat lower corolla lip with the stamens and style spanning over it and up toward the upper corolla lip. Bentham (1832) grouped the genus into three sections: *Ocimum* (*Ocymodon* Benth.), *Hierocymum* Benth., and *Gymnocymum* Benth. Further, on account of calyx morphology, section *Ocimum* was divided into subsections: *Ocimum*, *Gratissima*, and *Hiantia* Benth (Paton et al. 1999). Later, a different infrageneric classification was designed by Pushpangadan based on the morphological characteristics and divided the whole taxa into two groups: the Basilicum group and the Sanctum group. The species belonging to the Basilicum group were herbaceous annuals/perennials in nature having black, ellipsoid, strongly mucilaginous seeds, while the species belonging to the Sanctum group possessed of perennial shrubs with brown, globose, non-mucilaginous, or weakly mucilaginous seeds. According to Pushpangadan's system of classification, the Basilicum group contains only section *Ocimum* subsection *Ocimum* (Carvoic-Stanko et al. 2010). However, infrageneric classification designed by Pushpangadan does not act in congruence with the International Code of Botanical Nomenclature. Several cytotaxonomical, geographic distribution, classification and relationships, genetic diversity and phylogenetic, chemotaxonomical-based study have been carried out, but still often debated upon with regard to their nomenclature and taxonomical positioning. Keeping this in mind, our group had taken the challenge and reported comparative phylogenetic study of *Ocimum* species through inter-simple sequence repeat (ISSR) markers and non-coding plastid DNA region (Kumar et al. 2016).

Up till now, various molecular marker techniques, viz. random amplification of polymorphic DNA (RAPD), amplified fragment length polymorphism (AFLP), sequence-related amplified polymorphism (SRAP), ISSR, have been used for genetic analysis and characterization of the genus *Ocimum*. Singh et al. (2004) used RAPD for deducing phylogenetic relationship in five *Ocimum* species. According to study, *O. tenuiflorum* and *O. gratissimum* placed in one group and *O. basilicum*, *O. americanum*, and *O. kilimand scharicum* in another group, based on their genetic similarity. Similar type of results was also obtained by Carovic et al. (2006) and Patel et al. (2015) with RAPD. The position of *O. gratissimum* and *O. tenuiflorum* together in cluster could be explained by their belonging to the Sanctum group, while *O. basilicum*, *O. americanum*, *O. africanum*, and *O. kilimandscharicum* belong to the Basilicum group (Khosla 1995). PCR-based ISSR technique dominated over other marker techniques by virtue of its low cost, high sensitivity to low levels of genetic variation, and higher efficiency. Nowadays, ISSR–PCR is a very useful molecular tool for studying population genetics on a wide range of plant species, as well as for identifying species or cultivars. Some authors also worked on the characterization of the genus *Ocimum* with ISSR (Patel et al. 2015), AFLP (Carovic et al. 2006; Carovic-Stanko et al. 2011) and found same type of grouping as found with RAPD. High level of polymorphism and polymorphic information content depicted in the study reported by our group (Kumar et al. 2016) proved ISSR markers are appropriate for differentiating the genus *Ocimum*. However, Chen et al. (2013) while deducing genetic relationship in genus *Ocimum* from China using RAPD and ISSR found *O. gratissimum* clustered together with *O. americanum* and *O. basilicum*.

With the reference to the study done by our group, Neighbor joining tree depicted through ISSR-based analyses grouped the *Ocimum* spp. in three groups: "*Basilicum*", "*Sanctum*", and "*Gratissimum*" groups (Fig. 5.2a), which were in accordance with the earlier studies carried out by Paton et al. (1999). The grouping described was based on the morphological characters, which grouped the genus *Ocimum* into sections and subsections. Section "*Ocimum*" subsection "*Ocimum*" comprised of "*Basilicum*" group, and section "*Ocimum*" subsection "*Gratissima*" comprised of "*Gratissimum*" group, while *O. tenuiflorum* is placed in section "*Hierocymum*" subsection "*Foliosa*". Similar type of grouping was also reported by Carvoic-Stanko et al. (2010) based on molecular markers, nuclear DNA content, and chromosome number.

Studies based on the analysis of nucleotide sequences using standardized and universal DNA region/s have been also reported for phylogenetic analysis in *Ocimum*. Mattia et al. (2011) studied using *matK*, *rbcL*, and *trnH-psbA* and reported *O. tenuiflorum* and *O. gratissimum* grouped together in one cluster as *Sanctum* group. Similar type of result was also reported by Christina and Annamalai (2014) using *matK* and *rbcL*. However, Neighbor joining tree depicted through plastid *psbA-trnH* sequence dataset categorized the genus in two groups. Group I shows that the "*Gratissimum*" group was sister to "*Basilicum*" group, while, group II was comprised of *O. tenuiflorum* with its cultivars and *O. filamentosum* (Fig. 5.2b).

Various studies has been reported, which grouped "*Gratissimum*" group with "*Sanctum*" group on the basis of morphological, cytological, oil characters (Sobti and Pushpangadan 1979; Khosla 1995), genetic similarity defined through RAPD (Singh et al. 2004), and pollen morphology (Harley et al. 1992). This grouping can be explained on the fact that the ISSRs are distributed throughout the genome resulting in the amplification from both nuclear and organellar regions, while *psbA-trnH* region amplified only from cpDNA region. Chloroplast DNA (uniparental inheritance) possesses only half of the parentage of plants with hybridization/introgression or polyploidy and may insufficiently categorize plants into a group. Moreover, the size of nucleome is large when compared to the plastome and the chondriome. Hence, the origin of the ISSR bands is likely to be nuclear, and thus, we can conclude that the grouping with ISSR can be served as control.

5.4 Concluding Remark

During 2013–2017, in the Council of Scientific and Industrial Research (CSIR) supra network project ChemBio (BSC-0203), we came across several chemotypes and morphotypes as well as interspecific hybrid in genus *Ocimum*. Because of the high outcrossing pollination nature of species within the Basilicum group, they outcross with themselves leading to different chemotypes as well as morphotypes. Interestingly, self-pollinated species of *Ocimum,* viz. *O. gratissimum* and *O. tenuiflorum,* was also observed as different morphotypes in nature. The present article demonstrates brief reference to the researchers while working with the genus *Ocimum*. Moreover, in the

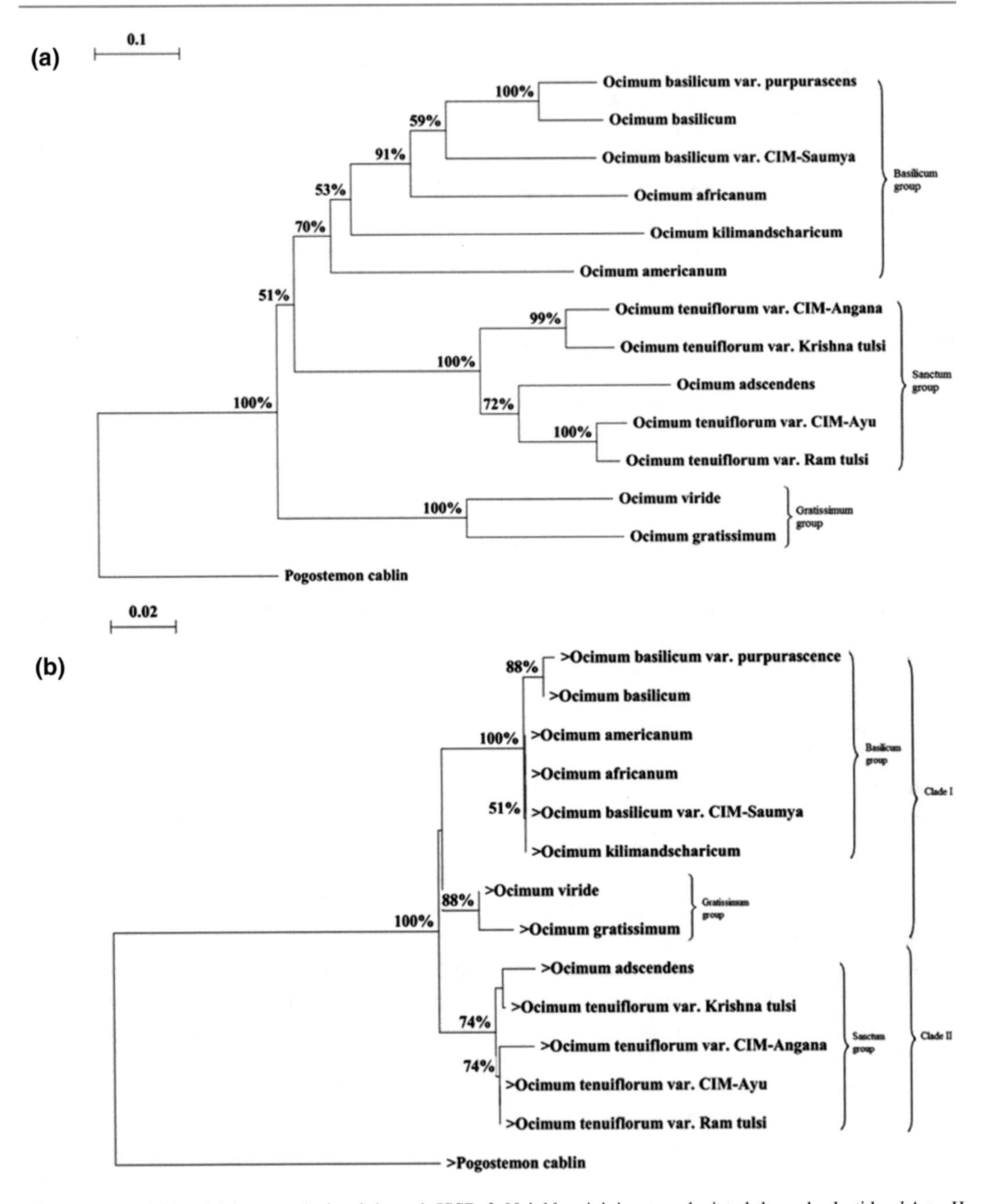

Fig. 5.2 **a** Neighbor joining tree depicted through ISSR. **b** Neighbor joining tree depicted through plastid *psbA-trnH* region (*Source* Kumar et al. 2016)

future, development of next-generation marker (DNA barcodes) for fast and accurate identification of the species is also prerequisite.

Acknowledgements Director, CSIR-CIMAP, Lucknow, is duly acknowledged for providing the laboratory facilities required to carry out this research. Authors also express their gratitude for the financial support from the Council of Scientific and Industrial Research (CSIR), New Delhi, through the XIIth FYP project ChemBio (BSC-0203).

References

Backer CA, Bakhuizen van den Brink RC (1965) Lamiaceae, Flora of Java. NVP Noordhoff, Groningen, pp 614–640

Balyan SS, Pushpangadan P (1988) A study on the taxonomical status and geographic distribution of genus *Ocimum*. Pafai J 10(2):13–19

Bentham G (1830) Synopsis of the Genera and Species of Indian Labiatae enumerated in the Catalogue of the Collection in Dr. Wallich's charge. In: Wallich N (ed) Plantae Asiaticae Rariores. Treuttel and Wurtz, London, pp 12–19

Bentham G (1832) Labiatarum Genera et Species 1. James Ridgway and Sons, London, pp 1–19

Bentham G (1848) Ocimoideae. In: de Candolle AP (ed) Prodromus Systematis Naturalis Regni Vegetabilis. Victor Masson, Paris, pp 30–148

Bentham G (1876) Ocimoideae. In: Bentham G, Hooker JD (eds) Genera Plantarum. Reeve and Co, London, pp 1162–1164, 1171–1179

Briquet J (1897) Labiatae. In: Engler A, Prantl K (eds) Die Natuirlichen Pflanzenfamilien. Wilhelm Engelmann, Leipzig, pp 183–375

Budantsev AL (1999) The Family Lamiaceae in Vietnam. Komarovia pp 3–33

Carovic K, Liber Z, Javornik B, Kolak I, Satovic Z (2006) Genetic relationships within Basil (*Ocimum*) as revealed by RAPD and AFLP markers. Acta Hortic 760:171

Carovic-Stanko K, Orlic S, Politeo O, Strikic F, Kolak I, Milos M, Satovic Z (2010) Composition and antibacterial activities of essential oils of seven *Ocimum* taxa. Food Chem 119:196–201

Carvoic-Stanko K, Liber Z, Besendorfer V, Javornik B, Bohanec B, Kolak I, Satovic Z (2010) Genetic relations among basil taxa (*Ocimum* L.) based on molecular markers, nuclear DNA content, and chromosome number. Plant Syst Evol 285:13–22

Carvoic-Stanko K, Liber Z, Politeo O, Strikic F, Kolak I, Milos M, Satovic Z (2011) Molecular and chemical characterization of the most widespread *Ocimum* species. Plant Syst Evol 294:253-262

Charles DJ, Simon JE (1992) A new geraniol chemotype of *Ocimum gratissimum* L. J Essent Oil Res 4:231–234

Chen SY, Dai TX, Chang YT, Wang SS, Ou SL, Chuang WL, Chuang CY, Lin YH, Lin YH, Ku HM (2013) Genetic diversity among *Ocimum* species based on ISSR, RAPD and SRAP markers. Aust J Crop Sci 7(10):1463–1471

Clement RA (1999) Labiatae. In: Long DG (ed) Flora of Bhutan. Royal Botanic Garden, Edinburgh, pp 938–1002

Cramer LH (1981) Lamiaceae (Labiatae). In: Dassanayake MD (ed) A revised handbook to the flora of ceylon. Oxford and IBH Publishing Co, New Delhi, pp 108–194

Doan T (1936) Labiatae. In: Lecomte H (ed) Flore Generale de L. Indo-Chine, Masson, Paris, pp 915–1046

Forsskål P (1775) Flora aegyptiaco-arabica. Moller, Kjobenhavn

Grayer RJ, Kite GC, Goldstone FJ, Bryan SE, Paton A, Putievsky E (1996) Intraspecific taxonomy and essential oil chemotypes in sweet basil, *Ocimum basilicum*. Phytochem 43:1033-1039

Gurke M (1895) Labiatae. In: Engler A (ed) Die Pflanzenwelt Ostafrikas. Teil C Dietrich Reimer, Berlin, pp 337–342

Hakkim FL, Arivazhagan G, Boopathy R (2008) Antioxidant property of selected *Ocimum* species and their secondary metabolite content. J Med Plants Res 2 (9):250–257

Harley MM, Paton A, Harley RM, Cade PG (1992) Pollen morphological studies in tribe *Ocimeae* (Nepetoideae: Labiatae): I. *Ocimum* L. Grana 31:161–176

Ho PH (1993) Labiatae. In: Ho PH (ed) An illustrated flora of Vietnam. Mekong Pring, Santa Ana, pp 1065–1108

Hooker JD (1885) Labiatae. Flora of British India. Reeve and Co, London, pp 604–705

Jacquin NJ (1792) Icones plantarum rariorum. Wappler CF, Vienna. White, London

Keng H (1969) Flora Malesianae precursors. A revision of Malesian Labiatae. Gard Bull, Singapore, pp 13–180

Keng H (1978) Labiatae. In: van Steenis CGGJ (ed) Flora Malesiana ser. Sijthoff and Noordhoff, Alpen Aan Den Rijn, pp 301–394

Keng H (1990) The concise flora of Singapore: gymnosperms and dicotyledons. Singapore University Press, Singapore

Khosla MK (1995) Study of inter-relationship, phylogeny and evolutionary tendencies in genus *Ocimum*. Indian J Genet Plant Breed 55:71–83

Kokwaro JO (1976) Medical Plants of East Africa. East African Literature Bureau, General Printers Limited, Nairobi, p 384

Kudo Y (1929) Labiatarum sino-japonicarum prodromus. Memoirs of the faculty of science and agriculture. Taihoku Imperial University, Taiwan, pp 37–332

Kumar A, Mishra P, Baskaran K, Shukla AK, Shasany AK, Sundaresan V (2016) Higher efficiency of ISSR markers over plastid *psbA-trnH* region in resolving taxonomical status of genus *Ocimum* L. Ecol Evol 6:7671–7682

Li XW (1977) Lamiaceae. In: Wu CY, Li HW (eds) Flora Reipublicae Popularis Sinicae. Academica Sinica, China

Li XW, Hedge I (1994) Labiatae. In: Wu ZY, Raven PH (eds) Flora of China. Missouri Botanical Garden, Missouri, pp 298–300

Linnaeus C (1753) Species Plantarum. Laurentii Salvii, Holmiae

Linnaeus C (1754) Genera Plantarum (ed). Laurentii Salvii, Holmiae

Loureiro J (1790) Flora Cochinchinensis. Ulyssipone, pp 357–375

Mattia FD, Bruni I, Galimberti A, Cattaneo F, Casiraghi M, Labra M (2011) A comparative study of different DNA barcoding markers for the identification of some members of Lamiacaea. Food Res Int 44:693–702

Merrill ED (1935) A commentary on Loureiro's "Flora cochinchinensis". Trans Am Phil Soc 1–445

Mukerjee SK (1940) A revision of the Labiatae of the Indian Empire. Rec Bot Surv India 14:1–228

Murata G (1971) Contributions to the Flora of Southeast Asia IV: a list of Labiatae known from Thailand. South E Asian Stud 8:489–517

Muschler R, Hosseus CC (1910) Beitrage zur Flora Siams. Beih Bot Centralbl 27:455–507

Obeng-Ofori D, Reichmuth CH, Bekele AJ, Hassanali A (1998) Toxicity and protectant potential of camphor, a major component of essential oil of *Ocimum kilimandscharicum*, against four stored product beetles. Int J Pest Manag 44:203–209

Patel HK, Fougat RS, Kumar S, Mistry JG, Kumar M (2015) Detection of genetic variation in *Ocimum* species using RAPD and ISSR markers. 3. Biotech 5:697–707

Paton AJ (1992) A synopsis of *Ocimum* L. in Africa. Kew Bull 47:405–437

Paton AJ, Putievsky E (1996) Taxonomic problems and cytotaxonomic relationships between and within varieties of *Ocimum basilicum* and related species (Labiatae). Kew Bull 51:509–524

Paton AJ, Harley RM, Harley MM (1999) *Ocimum*: an overview of classification and relationships. In: Hiltunen R, Holm Y (eds) *Ocimum*. medicinal and aromatic plants—industrial profiles. Harwood Academic, Amsterdam, pp 1–38

Phuong VX (1982) Conspectus Lamiacearum Florae Vietnam. Novosti Sist Nizsh Rast 19:125–160

Phuong VX (1995) The Lamiaceae in the Flora of Vietnam. J Biol 17:33–46

Phuong VX (2000) Lamiaceae (Labiatae). In: Ban NT, Ly TD, Khanh TC, Loc PK, Thin NN, Tien NV, Khoi NK (eds) Flora of Vietnam. Science and Technics Publishing House, Hanoi, pp 1–278

Pino JA, Rosado A, Fuentes V (1996) Composition of the essential oil from the leaves and flowers of *Ocimum gratissimum* L. grown in Cuba. J Essen Oil Res 8:139–141

Prasad G, Kumar A, Singh AK, Bhattacharya AK, Singh K, Sharma VD (1986) Antimicrobial activity of essential oils of some *Ocimum* species and clove oil. Fitoterapia 57:429–432

Press JR (1982) Elsholtzia. In: Hara H, Chater AO, Williams LHJ (eds) An enumeration of the flowering plants of Nepal. British Museum (Natural History), London

Pushpangadan P, Bradu BL (1995) Medicinal and aromatic plants. In: Chadha KL, Gupta R (eds) Advances in horticulture. Malhotra Publishing House, New Delhi, pp 627–657

Rastogi S, Kalra A, Gupta V, Khan F, Lal RK, Tripathi AK, Parameswaran S, Gopalakrishnan C, Ramaswamy G, Shasany AK (2015) Unravelling the genome of Holy basil: an "incomparable" "elixir of life" of traditional Indian medicine. BMC Genom 16:413

Ridley HN (1923) Labiatae. In: Ridley HN (ed) The flora of the Malay Peninsula. Reeve and Co, London, pp 624–655

Simon JE, Quinn J, Murray RG (1990) Basil: a source of essential oils. In: Janick J, Simon JE (eds) Advances in new crops. Timber Press, Portland, pp 484–989

Singh RK, Dwivedi RS (1987) Effect of oils on *Sclerotium rolfsii* causing foot rot of barley. Indian Phytopathol 40:531–533

Singh AP, Dwivedi S, Bharti S, Srivastava A, Singh V, Khanuja SPS (2004) Phylogenetic relationships as in *Ocimum* revealed by RAPD markers. Euphytica 136:11–20

Sobti SN, Pushpangadan P (1979) Cytotaxonomical studies in the genus *Ocimum*. In: Bir SS (ed) Taxonomy, cytogenetics and cytotaxonomy of plants. Kalyani Publication, New Delhi, pp 373–377

Suddee S, Paton AJ, Parnell JAN (2005) Taxonomic revision of tribe *Ocimeae* Dumort. (Lamiaceae) in Continental South East Asia III. Ociminae. Kew Bull 60:3–75

Tadros MG, Ezzat SM, Salama MM, Farag MA (2014) In vitro and in vivo Anticholinesterase activity of the volatile oil of the aerial parts of *Ocimum basilicum* L. and *Ocimum africanum* Lour. Growing in Egypt. Int J Med Health Biomed Bioeng Pharm Eng 8(3):157–161

Tokin B (1944) Phytoncides or plant bactericides, upon protozoa. Am Rev Soviet Med 1:237–239

Weaver DK, Dunkel FV, Ntezurubanza L, Jackson LL, Stock DT (1991) The efficacy of linalool, a major component of freshly-milled *Ocimum canum* Sims (Lamiaceae), for protection against postharvest damage by certain stored product Coleoptera. J Stored Prod Res 27:213–220

Verma RS, Kumar A, Mishra P, Baskaran K, Padalia RC, Sundaresan V (2016) Essential oil composition of four *Ocimum* spp. from the Peninsular India. J Essen Oil Res 28:35–41

Vina A, Murillo E (2003) Essential oil composition from twelve varieties of basil (*Ocimum* spp.) grown in Colombia. J Braz Chem Soc 14:744–749

Genetics, Cytogenetics, and Genetic Diversity in the Genus *Ocimum*

6

Soni Gupta, Abhilasha Srivastava, Ajit Kumar Shasany and Anil Kumar Gupta

Abstract

The genus *Ocimum* of the family Lamiaceae has been known since centuries primarily for its essential oil, which is used extensively in pharmaceutical, culinary, and perfumery industries. In addition to the wild *Ocimum* spp., commercially cultivated varieties, popularly known as basils, are found in different geographical regions of the world. The genus is well known for its morphological and chemical diversity. This widespread variability is attributed to inter- and intraspecific hybridization, polyploidy, aneuploidy, synonymous names, various varieties and cultivars, and huge plethora of chemotypes. These have resulted in ambiguities in classification of the genus and consequently a complex taxonomic scenario. Initially, basils were identified according to their geographic origin that had distinct chemotypes in the world market. Over the years, morphological, chemical, karyological, and molecular means are being applied to get a better understanding of the genetic relationship among basils. The base chromosome numbers suggested are $x = 12$ (*O. basilicum* clade as tetraploid; *O. americanum* clade as hexaploids), $x = 10$ (*O. gratissimum*), and $x = 9$ (*O. tenuiflorum*). Variations in genome size and chromosome number among *Ocimum* spp. imply that sequence deletion/amplification, chromosome rearrangements, and polyploidization have played a role in the course of evolution. Recent revision by nuclear DNA content has divided the section *Ocimum* into two clades—the first comprising *O. basilicum* and *O. minimum,* whereas the second comprising *O. americanum*, *O. africanum*, and two *O. basilicum* var. *purpurascens* accessions. *O. tenuiflorum* was found to be the most divergent species. It was also seen that *O. basilicum* clade species are tetraploids, while species belonging to *O. americanum* clade are hexaploids. For the first time, DNA linkage map for sweet basil has been constructed, which is anchored by SSRs (42 EST-SSR) and saturated by SNPs (1847) spanning 3030.9 cM and QTLs identified for basil downy mildew response. The plant

S. Gupta · A. Srivastava · A. K. Gupta (✉)
Genetics and Plant Breeding Division, CSIR-Central Institute of Medicinal and Aromatic Plants,
Lucknow 226015, India
e-mail: ak.gupta@cimap.res.in

S. Gupta
e-mail: sonimail2u@gmail.com

A. Srivastava
e-mail: abhilasha.cimap@gmail.com

A. K. Shasany
Biotechnology Division, CSIR-Central Institute of Medicinal and Aromatic Plants,
Lucknow 226015, India
e-mail: ak.shasany@cimap.res.in

A. K. Shasany and C. Kole (eds.), *The Ocimum Genome*, Compendium of Plant Genomes, https://doi.org/10.1007/978-3-319-97430-9_6

genetic resources existing in *Ocimum* have been taken up for diversity studies and constitute the much required raw materials for future breeding programs for desired chemotypes, disease resistance, and better agronomic traits.

6.1 Introduction

Ocimum, the largest genus of the family Lamiaceae, is native to tropical and warm temperate areas and is represented by aromatic annual and perennial shrubs with the greatest number of species in the African continent (Mukherjee and Dutta 2007, Chowdhury et al. 2017). The genus *Ocimum* comprises more than 150 species; however, the most popular species are *Ocimum basilicum, O. americanum* (syn. *O. canum*), *O. gratissimum, O. kilimandscharicum, O. tenuiflorum* (syn. *O. sanctum*), and *O.* × *citriodorum* (syn. *O.* × *africanum*), which is a hybrid between *O. basilicum* and *O. americanum* (Paton et al. 1999; Carović-Stanko et al. 2010; Moghaddam et al. 2011; Rewers and Jędrzejczyk 2016).

The highly aromatic plants of the genus, popularly known as basil, have long been established as an economically important herb due to their essential oils (monoterpenes, sesquiterpenes, and phenylpropanoids as constituents) that have medicinal, culinary, and perfumery applications. The leaves and flowering tops of the plant bear specialized structures called peltate trichomes that are the site for essential oil biosynthesis and accumulation. The oil is extracted after steam distillation of the aerial parts (Werker 1993). Biologically active constituents in the essential oil are found to possess antioxidant, insecticidal, nematicidal, antimicrobial, and anticancerous properties (Datta et al. 2010; Patel et al. 2015). Important essential oil metabolites reported from *Ocimum* species include linalool, linalyl, geraniol, citral, camphor, eugenol, methyl eugenol, methyl chavicol, methyl cinnamate, thymol, safrol, taxol, ursolic acid (Upadhyay et al. 2015).

The oil composition is seen to vary according to geographical locations, and four different types are recognized and reported on the basis of oil profile—(i) European type (sweet basil) having superior aroma quality contains linalool and methyl chavicol as major components and no camphor; it is similar to Egyptian basil; (ii) reunion type (reported originally from Comoro Islands and also from Madagascar, Thailand, and Vietnam) is characterized by high concentrations of methyl chavicol; (iii) methyl cinnamate type (reported from Bulgaria, India, Guatemala, and Pakistan) contains methyl chavicol, linalool, and methyl cinnamate; and (iv) eugenol type (reported from Java, Russia, and North Africa) contains eugenol as major component (Wealth of India 1976; Heath 1981; Sobti et al. 1982; Simon et al. 1990; Marotti et al. 1996; Datta et al. 2010).

In India, the genus is represented by only nine species including *O. adscendens*, *O. basilicum, O. canum, O. gratissimum, O. kilimandscharicum, and O. tenuiflorum* and three exotic species, namely *O. americanum* L., *O. minimum* L., and *O. africanum* Lour (Balyan and Pushpangadan 1988). *O. tenuiflorum* (earlier known as *O. sanctum*) is commonly known as 'Tulsi' meaning 'matchless one' in Sanskrit and has been described as holy and medicinally important plant in ancient scriptures. In Ayurveda, it is documented as 'elixir of life' and is claimed for longevity (Puri 2002).

6.2 Taxonomic Position

A vague understanding of genetic relationship among basils has resulted in a situation of taxonomic confusion hitherto unsolved (Harley et al. 1977; Grayer et al. 1996). This is due to prevalent interspecific hybridization, polyploidy, and aneuploidy, numerous botanical varieties, cultivar names, and a number of morphological similar but distinct chemotypes present in this genus (Simon et al. 1990). Further, human interference in terms of selection and hybridization along with synonymous names adds to this confusion (Putievsky et al. 1999; Labra et al. 2004; Mukherjee et al. 2005; Moghaddam et al. 2011).

Linnaeus (1753) listed five species in *Ocimum* that was further expanded by Bentham in 1832 up to 40 species which were grouped into three sections: *Ocimum* (Ocymodon Benth.), *Hierocymum* Benth., and *Gymnocymum* Benth. (Paton et al. 1999; Carović-Stanko et al. 2010). The section *Ocimum* was further divided into three subsections: *Ocimum, Gratissima*, and *Hiantia* Benth. A new section, Hemizygia Benth., was added by Bentham which was recognized by Briquet (1897) as a separate genus due to fused anterior stamens. Paton (1992) used infrageneric classification of *Ocimum* proposed by Bentham (1948) and removed Hemizygia and subsect. Hiantia, giving the latter as a separate status of genus *Becium* (Paton et al. 1999). Pushpangadan (1974) proposed a different infrageneric classification.

On the basis of morphological and cytological characters, Sobti and Pushpangadan (1977) classified Indian species of *Ocimum* into two broad categories—Basilicum (*O. basilicum, O. canum, O. kilimandscharicum*) and Sanctum (*O. tenuiflorum and O. gratissimum*). The Basilicum group includes mostly annual herbs with stalked bracts, conspicuous flowers, black mucilaginous seeds, and with chromosome number varying from $2n = 24$ to 76 with basic number as $x = 12$. On the other hand, the Sanctum group is represented by perennial or under-shrubs having sessile bracts, inconspicuous flowers, and brownish shiny globose or subglobose with non-mucilaginous seeds and with chromosome numbers varying from $2n = 32$ to 40 with basic chromosome number $x = 8$ (Datta et al. 2010).

According to Pushpangadan's classification (1974), the Basilicum group consists of the only section *Ocimum* subsection *Ocimum*. Six important botanical varieties and cultivars dominating the world market (var. *basilicum* cv. Genovese, var. *basilicum* cv. sweet basil, var. *difforme* Benth., var. *purpurascens* Benth., v. Dark Opal, and var. *thyrsiflorum/L./* Benth.) were placed in the Basilicum group, section *Ocimum* (=*O. basilicum* group) by Pushpangadan. However, it was seen that the infrageneric classification was not in accordance with the International Code of Botanical Nomenclature (Carović-Stanko et al. 2010).

Efforts to revise the genus have been attempted at the Royal Botanical Garden, Kew, London, (Paton 1992) and also by several taxonomists around the world on the basis of geographical origin, morphological, karyological, essential oil composition, crossability, and molecular means (Khosla 1995; Grayer et al. 1996; Marotti et al. 1996; Paton and Putievsky 1996; Ravid et al. 1997; Martins et al. 1999; Putievsky et al. 1999; Erum et al. 2011; Malav et al. 2015). Paton and Putievsky (1996) proposed a system of standardized descriptors including the volatile oil for efficient communication and identification in *Ocimum*. *Ocimum* is extensively cultivated, and natural and/or induced cross-hybridizations lead to varying ploidy levels, newer species, subspecies, and varieties that may or may not be morphologically distinct (Chowdhury et al. 2017). On the other hand, *Ocimum* species show huge morphological and chemical variations along with growth- and reproductive-related traits among themselves, which are additionally affected by environmental factors (Carović-Stanko et al. 2010). Also, genetic variability at inter- and intraspecies level is also seen. Thus, taxonomically morphological descriptors create a lot of confusion (Chowdhury et al. 2017).

Lawrence (1992) and Grayer et al. (1996) exploited chemotaxonomy in classifying *Ocimum* on the basis of chemotypes which basically uses the most abundant volatile component of the essential oil. The chemical composition varies not only among species but also cultivars, which can be further affected by cross-hybridization, morphogenesis, polyploidization, process of oil extraction, drying and storage, stages of harvesting, etc. (Chowdhury et al. 2017). Thus, chemotaxonomy further aggravates the taxonomic confusion. So, it has been suggested that conventional taxonomic evaluation and chemotaxonomy have to be taken together with other molecular tools for classification (Labra et al. 2004; Carović-Stanko et al. 2011).

6.3 Cytogenetics of *Ocimum*

The base chromosome number for the genus *Ocimum* is $x = 8$ according to Darlington and Wylie (1955) and Mehra and Gill (1972).

Two base numbers $x = 8$, 12 were suggested by Morton (1962) in *Ocimum*. Sobti and Pushpangadan (1977) regarded $x = 12$ as the predominant base number for 'Basilicum group' including the species *O. canum* ($2n = 24$, 26), *O. basilicum* ($2n = 48$), *O. americanum* ($2n = 72$), and *O. kilimandscharicum* ($2n = 72$) and $x = 8$ as the most common base number for 'Sanctum group' in genus *Ocimum*. Evolution of new basic chromosome numbers accompanied by aneuploidy and polyploidy has played a major role in the diversification of the species in *Ocimum* (Carović-Stanko et al. 2010). A wide range of chromosome numbers was reported in Sanctum group ($n = 16$, 19, 20, 24, and 34) (Vij and Kashyap 1976; de-Wet 1958; Morton 1962; Mitra and Datta 1967; Mehra and Gill 1972; Getsadze 1975; Singh 1978). Khosla (1989) proposed that the haploid chromosome number $x = 8$ of 'Sanctum group' and other higher numbers in different members probably have evolved from a primitive base number $x = 6$ through dysploidy and polyploidy at different levels.

Similarly, reports of different chromosome numbers in the same species have been documented such as in *O. tenuiflorum* ($2n = 32$, 36, and 76) and *O. minimum* ($2n = 48$ and 56) (Khosla 1995; Paton and Putievsky 1996; Mukherjee and Datta 2006). Mukherjee et al. (2005) suggested, based on secondary chromosome association in meiosis, that cytological diploidization of some *Ocimum* species (*O. basilicum*, *O. canum*, *O. kilimandscharicum*, *O. gratissimum*, *O. tenuiflorum*) during evolution resulted in functional diploids with $x = 12$, which probably evolved from $x = 6$.

Carović-Stanko et al. (2010) suggested $x = 12$ (*O. basilicum* clade as tetraploid; *O. americanum* clade as hexaploids), $x = 10$ (*O. gratissimum*), and $x = 9$ (*O. tenuiflorum*) as base chromosome numbers. Variations in genome size and chromosome number among *Ocimum* spp. imply the role of chromosomal rearrangements, sequence deletion/amplification, and polyploidization in the course of evolution.

Omidbaigi et al. (2010) induced autotetraploids by colchicine treatment in *O. basilicum* and found larger stomata, pollen grains, more chloroplasts in guard cells, and less stomatal density as compared to diploid control plants.

6.4 Genetic Diversity

Ocimum grows in all the six inhabited continents and is extensively cultivated in India, Iran, Japan, China, and Turkey (Sadeghi et al. 2009). As mentioned earlier, abundant variability in morphological and chemical traits exists in nature due to polyploidy, aneuploidy, and inter- and intraspecific hybridizations in addition to targeted cultivation and breeding practices for desired morpho-chemotypes. Additionally, several biotic and abiotic factors affect the level of essential oil accumulation and its components (Bernhardt et al. 2015). The phenotypic and chemotypic variations manifested in several *Ocimum* species across various geographical regions have been extensively studied and documented by researchers by several means, viz. morphological, chemical, karyological, and molecular. Some are elaborated below.

6.4.1 Morpho-Chemotypes

Commercially important basil cultivars mostly belong to the species *O. basilicum*. Darrah (1980) classified the *O. basilicum* cultivars into seven categories: (1) tall slender types, (the sweet basil group); (2) large-leafed robust types ('lettuce leaf' also called 'Italian' basil); (3) dwarf types, which are short and small leafed ('bush' basil); (4) compact types, also described *O. basilicum* var. *thyrsiflora* ('Thai' basil); (5) *purpurascens*, the purple-colored basil types with traditional sweet basil flavor; (6) purple types ('Dark Opal': hybrid between *O. basilicum* and *O. forskolei* with a sweet basil plus clove-like aroma); and (7) *citriodorum* types (lemon-flavored basils). Bernhardt et al. (2015) discussed several factors such as genetic background, ontogenesis, morphogenesis, essential oil extraction method, drying, and storage that affect the level of essential oil accumulation and its components in *Ocimum* species. Egata et al.

(2017) recorded the morpho-agronomic variability to further select promising sweet basil accessions in Ethiopia. Singh et al. (2018) examined the genetic diversity and clustering pattern among 20 five basil accessions and found oil content was the highest contributing character toward the genetic diversity (56.09%). Srivastava et al. (2018) studied 60 *O. basilicum* accessions and found some unique chemotypes which could be further exploited in future breeding programs. Carović-Stanko et al. (2011) studied the resolving power of morphological traits for reliable identification of *O. basilicum* accessions and categorized six clusters of basil morphotypes. Standardized descriptor list enlisting morphological traits developed by the International Union for the Protection of New Varieties of Plants (UPOV) (2003) was used, and it was concluded that stringent selection of morphological traits and cautious analysis can help in routinely screening and managing germplasm accessions inexpensively and reliably. For morphological characterization, the descriptor list containing several plant traits related to habit, leaf, inflorescence, flower, seed, oil, etc., has been developed by National Bureau of Plant Genetic Resources (NBPGR), India (NBPGR 2002).

6.4.2 Nuclear DNA Content

Carović-Stanko et al. (2010) for the first time reported nuclear DNA content on *Ocimum* species, and their results supported the existence of more infrageneric groups within the genus. They divided the section *Ocimum* into two clades—the first included *O. basilicum* and *O. minimum,* whereas the second comprised *O. americanum*, *O. africanum*, and two *O. basilicum* var. *purpurascens* accessions. *O. tenuiflorum* was found to be the most divergent species with the smallest genome size, organized in small chromosomes, and the least chromosome number. The study implied that the species belonging to *O. basilicum* clade are tetraploids, while species belonging to *O. americanum* clade are hexaploids. Rewers and Jedrzejczyk (2016) found that *Ocimum* species possessed very small, small, and intermediate genomes. Additionally, different values of genome size occurred within one species implying the presence of polyploids and aneuploids. Three different values of genome size were detected in *O. americanum* accessions (2.3, 4.4, and 7.4 pg/2C), whereas *O. basilicum* possessed genome size with 4.4–4.9 pg/2C DNA content. One Indian accession and another Iranian accession possessed a genome size of 7.3 and 7.4 pg/2C. Very small genomes were detected in *O. campechianum* (12 pg/2C) and in *O. gratissimum* (1.8–2.3 pg/2C), while small genome in *O. selloi* (3.1 pg/2C). *O. tenuiflorum* accessions had 0.9, 1.9, and 4.5 pg/2C DNA content. They also did characterization using inter-simple sequence repeats (ISSRs) markers in addition to flow cytometry.

6.4.3 Molecular Markers

Due to complex relationships between *Ocimum* species and huge intraspecific variability in morphological and biochemical traits, the morphological differentiation of these species is very difficult. So, a combination of morphological, karyotypic, chemical, and molecular markers for an unambiguous conclusion about cultivated basils is required (Wetzel et al. 2008).

Molecular diversity in *Ocimum* has been studied by several workers employing random amplified polymorphic DNA (RAPD), inter-simple sequence repeat (ISSR), amplified fragment length polymorphism (AFLP), sequence-related amplified polymorphism (SRAP), simple sequence repeat (SSR), internal transcribed spacer (ITS) (Labra et al. 2004; De Masi et al. 2006; Carović-Stanko et al. 2010, 2011; Shinde et al. 2010; Moghaddam et al. 2011; Aghaei et al. 2012; Chen et al. 2013; Mahajan et al. 2015; Malav et al. 2015; Chowdhury et al. 2017). These have been studied along with the morpho-chemical traits of the *Ocimum* accessions.

6.4.4 Isoenzymes

Amaral et al. (2000) identified and characterized the isoenzymatic variability in two populations: (a) the Nova Friburgo-RJ and (b) the Tiradentes-MG of *Ocimum selloi* Benth. by using several enzymatic systems (esterase, acid phosphatase, glutamate dehydrogenase, glutamate oxaloacetate, transaminase, leucine aminopeptidase, peroxidase, and shikimate dehydrogenase). Extraction and characterization of peroxidase from *Ocimum tenuiflorum* and *Ocimum gratissimum* have also been attempted by Anbuselvi et al. (2013). Activity of an isozyme alpha esterase in different tissues of *Ocimum sanctum* was also reported by Padmanaban et al. (2013).

6.4.5 DNA Barcodes

Recently, DNA barcodes also have been used to identify particular species in the genus (De Mattia et al. 2011; Bast et al. 2014; Christina and Annamalai 2014; Bhamra et al. 2014; Elansary et al. 2015). Kumar et al. (2016) reported on the comparative phylogenetic study of *Ocimum* species through ISSR markers and plastid DNA region.

6.5 Inheritance of Traits

An insight into the study of heritability and genetic advance helps in estimating the nature of inheritance of a trait. Heritability reflects the proportion of genotypic variability which is transmitted from parents to progeny. It helps in the selection of elite genotypes from a variable population. However, high heritability coupled with genetic advance serves better in the ultimate selection process.

Khan et al. (2012) reported high estimates of heritability in *Ocimum* spp. for plant height (99.70%), inflorescence/plant (98.80%), fresh herb yield per plant (98.60%), number of inflorescence/plant (98.40%), days to maturity (92.60%), and leaf width (91.50%). Moderate heritability was recorded for number of primary branches/plant (73.20%), number of secondary branches/plant (68.00%), leaf length (59.50%), and leaf width (76.70%). In a study on *Ocimum gratissimum*, Edet (2018) reported high values of heritability and genetic advance for characters like plant height, leaf length, leaf width, leaf area, and fresh weight. A moderate value was recorded for raceme length and dry weight. Ibrahim et al. (2013) carried out a study consecutively for two growing seasons on three varieties, i.e., French, Purple, and Lemon basils, and concluded that heritability and expected genetic advance had the highest values in case of herb dry weight, stem dry weight, leaf dry weight, and linear growth, respectively. The lowest values were recorded for essential oil content and number of primary branches. Ahmad and Khaliq (2002), in a study on *Ocimum sanctum*, also reported a high level of heritability for plant height, leaf area, 1000-seed weight, number of racemes/plot, and number of flowers per raceme. In a study on segregating generations of basil by Singh et al. (2013), the estimates of heritability for the majority of traits in all F_2, BC_1, BC_2 were adequately higher for a number of primary branches and oil content. The values of genetic advance in F_2 population were higher than that in other population for fresh herb yield, dry herb yield, and oil content.

This variation in the highest and lowest values of heritability is a result of the environmental effect. The traits with high heritability are less influenced by the environment. It shows that the influence of genetic effect was more than the environmental effect. Also, such high levels of heritability and genetic advance are the result of the additive gene action involved in the inheritance of these traits. Thus, the selection for improvement of such traits in further generations would be rewarding.

6.6 Linkage Map

Molecular genetic linkage mapping and DNA markers linked to important traits controlled by quantitative trait loci (QTLs) are gaining importance in modern plant breeding programs.

Pyne et al. (2017) developed expressed sequence tag SSR (EST-SSR) and single-nucleotide polymorphism (SNP) markers from double digestion restriction site-associated DNA sequencing (ddRADseq) and used to genotype the MRI x SB22 F_2 mapping population, which segregated for response to downy mildew (*Peronospora belbahrii*) in sweet basil. They constructed the first linkage map for sweet basil, which was anchored by SSRs (42 EST-SSR) and saturated by SNPs (1847) spanning 3030.9 cM (Fig. 6.1). They identified three QTLs model that explained 37–55% of phenotypic variance associated with downy mildew response (Table 6.1). A single major QTL, *dm11.1*, explained 21–28% of phenotypic variance and demonstrated dominant gene action. Two minor QTLs, *dm9.1* and *dm14.1*, explained 5–16% and 4–18% of phenotypic variance, respectively.

6.7 Plant Genetic Resources

Agro-biodiversity essentially comprises plant genetic resources (PGRs) as its main component (Upadhyaya et al. 2008). Traditionally utilized plant species, landraces, modern and obsolete cultivars, breeding lines, weedy types, and wild species comprise the PGR (IPGRI 1993). They serve as the genetic material for the crop improvement programs. It is a well-known fact that these PGRs are finite and prone to losses due to several factors such as cultivation practices, urbanization and environmental catastrophes. Over the years, several countries have contributed toward the conservation of these PGRs in world over 1750 gene banks housing more than 7.5 million gene bank accessions (FAO 2013) with the goal of achieving food security and reducing poverty in developing countries by R&D efforts in the fields of agriculture, forestry, fisheries, policy, and environment (Upadhyaya et al. 2008).

Aligning to the global effort to conserve the biodiversity, the National Gene Bank at Council of Scientific and Industrial Research—Central Institute of Medicinal and Aromatic Plants (CSIR-CIMAP), Lucknow, houses several exotic and Indian *Ocimum* germplasm accessions. These comprise a total of five species of *Ocimum* (*O. sanctum*, *O. basilicum*, *O. africanum*, *O. gratissimum*, and *O. kilimandscharicum*). Out of these five species, *O. kilimandscharicum* and *O. gratissimum* are perennial species while the other three are annual.

Qualitative and quantitative morphological and chemical traits as described in the list of descriptors were used to catalogue and characterize 80 accessions of *O. basilicum* from different geographical regions (India, Singapore, Tanzania, Thailand, and Slovak Republic). The morphological characters revealed a wide range of variation among themselves, viz. plant height (56–126 cm), average number of primary branches (4–8), internode length (2.5–7.9 cm), inflorescence length (16–25 cm), leaf area (1.81–8.11 cm), herb yield per plant (<500–3125 gm), oil content (0.2–0.98%), oil yield per plant (0.5–12.21 ml per plant). Figure 6.2 depicts the percent accessions in the germplasm which show variation from the normal range for each of the trait under study.

Further, Srivastava et al. (2018) estimated the genetic and chemotypic variability among the subset of the above accessions (Table 6.2) and statistically grouped into seven diverse clusters by Mahalanobis D^2 analysis. One of the accessions, OB 50, was found to be most divergent from rest of the accessions. The main character responsible for its uniqueness was found to be herb yield per plant.

There was no relation between the genetic diversity and geographical place of origin as the accessions of different origins including the exotic accessions were grouped into the same cluster.

The contribution of characters toward the total divergence helps in the selection and choice of parents. Among the nine characters studied, leaf area (16.01%) contributed the maximum toward the genetic diversity followed by oil yield/plant (15.12%) and internodal distance (14.58%). Therefore, while doing selections for further improvement, these characters can be given due importance.

The chemical diversity of the essential oil of the above accessions grouped them into two broad clusters (A and B) on the basis of the

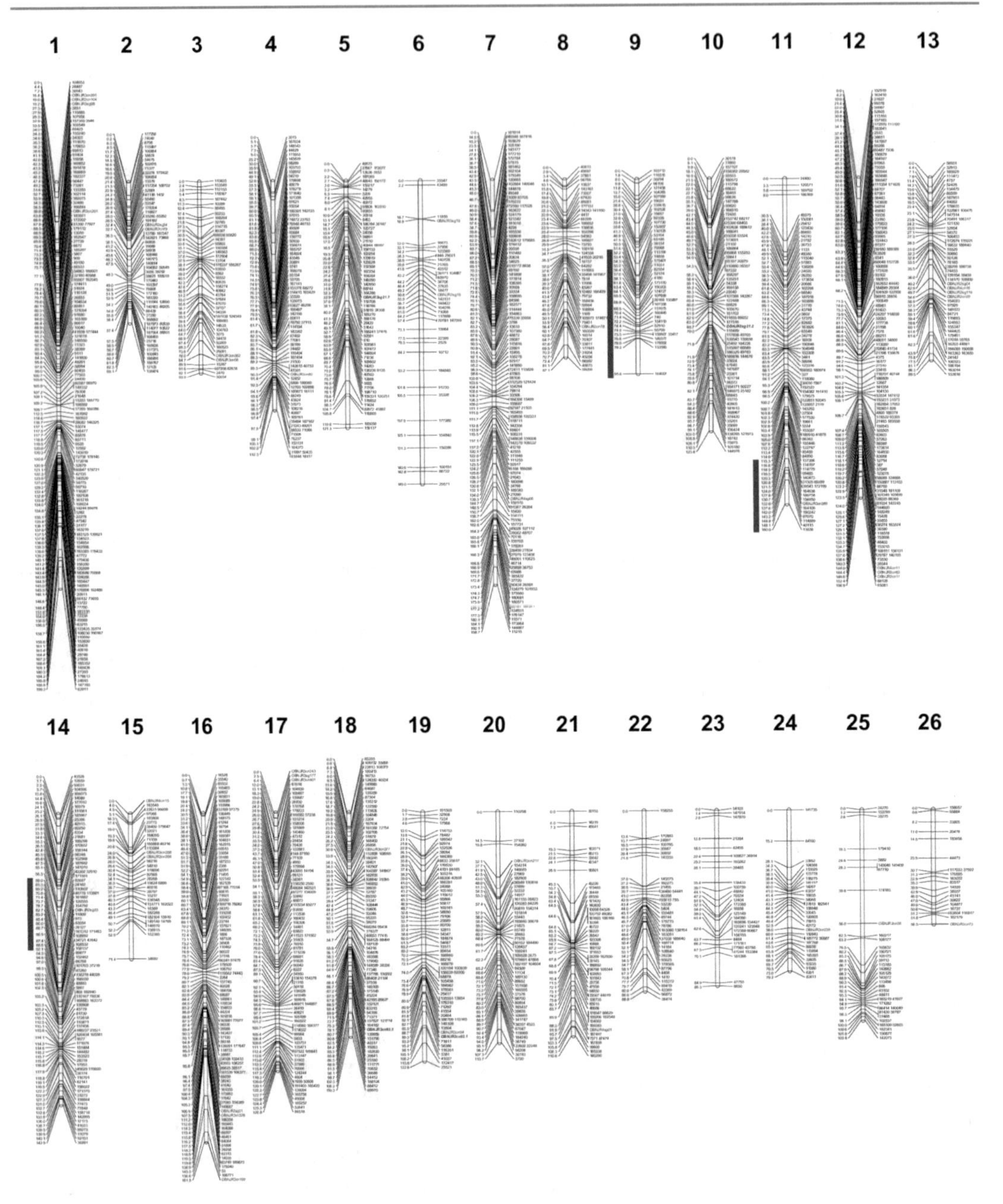

Fig. 6.1 Sweet basil linkage map constructed for MRI × SB22 F2 intercross family. Pyne et al. (2017) A first linkage map and downy mildew resistance QTL discovery for sweet basil (*Ocimum basilicum*) facilitated by double digestion restriction site-associated DNA sequencing (ddRADseq). PLOS ONE 12(9): e0184319. https://doi.org/10.1371/journal.pone.0184319

dominance of either methyl chavicol or linalool in its oil. Cluster A had seven accessions highly rich in methyl chavicol ($\geq 90\%$), ten accessions rich in methyl chavicol along with methyl eugenol, and twenty-five accessions rich in methyl chavicol along with linalool (Fig. 6.3). In Cluster B, five lines were found to be highly rich in linalool while some accessions had linalool along with methyl cinnamate. One of the accessions, OB 28, was rich in both methyl cinnamate

Table 6.1 Summary of three downy mildew resistance QTL detected using a multiple QTL model (MQM) across three environments

QTL	LG	Position (cM)	SNP[a]	Confidence Interval (cM)[b]	Environment	LOD	P[c]	PVE (%)[d]
dm 9.1	9	74.9	95799	51.9–95.6	NJSN14		<0.001	16.1
		–	–	–	NJRA14	2.1	0.012	6.5
		–	–	–	NJRA15	1.5	0.047	5.5
dm 11.1	11	160.0	11636	115.3–160.0	NJSN14	7.2	<0.001	20.6
		160.0	11636	115.3–160 0	NJRA14	6.7	<0.001	23.3
		160.0	11636	114.0–160.0	NJRA15	6.5	<0.001	28.2
dm 14.1	14	73.7	120555	65.3–92.1	NJSN14	6.6	<0.001	18.4
		73.7	120555	65.3–131.0	NJRA14	3.3	<0.001	10.5
		–	–	–	NJRA15	1.1	0.109	3.9

Pyne et al. (2017) A first linkage map and downy mildew resistance QTL discovery for sweet basil (*Ocimum basilicum*) facilitated by double digestion restriction site-associated DNA sequencing (ddRADseq). PLOS ONE 12(9): e0184319. https://doi.org/10.1371/journal.pone.0184319

[a]Single-nucleotide polymorphism (SNP) marker located in closest proximity to the QTL location

[b]1.5 LOD score intervals shown or significant ($P < 0.01$) QTL only

[c]*P* values represent the significance of LOD scores determined by permutation tests with 1000 iterations at $\alpha = 0.05$

[d]Percent phenotypic variance explained

https://doi.org/10.1371/journal.pone.0184319.t003

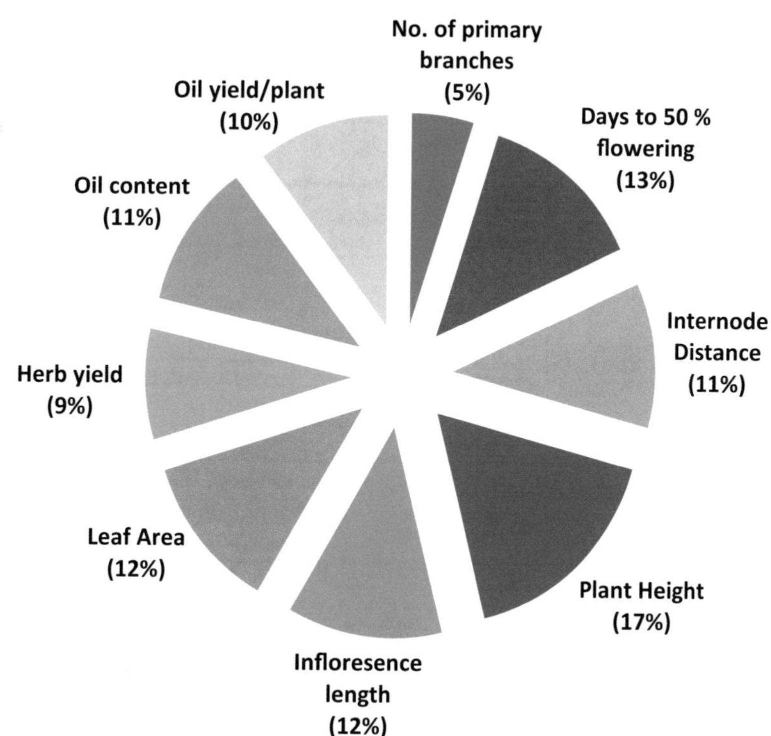

Fig. 6.2 Morphological variability in *Ocimum* germplasm at CSIR-CIMAP, Lucknow. The numbers in brackets show the percentage of accessions showing deviation from the normal range for the given trait

Table 6.2 *O. basilicum* germplasm accessions for genetic diversity study

S. No.	Accession No.	Accessions/cultivar	Place of collection/origin	Distinct features
1	OB-1	French basil	Chennai, AP (India)	
2	OB-2	Vikarsudha	CSIR-CIMAP, Lucknow UP (India)	
3	OB-3	Sweet basil	Gandhi Nagar, Gujarat (India)	
4	OB-4	French basil	Bangalore, Karnataka (India)	
5	OB-5	French basil	Mangalore, Karnataka (India)	
6	OB-6	French basil	Chandigarh	
7	OB-7	Sweet basil	CSIR-CIMAP, Lucknow UP (India)	
8	OB-8	Sweet basil	Singapore	
9	OB-9	Sweet basil	Singapore	
10	OB-10	Sweet basil	CSIR-CIMAP, Lucknow UP (India)	
11	OB-11	Sweet basil	Košice, Slovak Republic	
12	OB-12	Sweet basil	CSIR-CIMAP, Lucknow UP (India)	
13	OB-13	French basil	Mangalore, Karnataka (India)	Highly rich in methyl chavicol
14	OB-14	Indian basil	Muzaffarpur, Bihar (India)	
15	OB-15	Indian basil (CIM-Saumya)	CSIR-CIMAP, Lucknow UP (India)	
16	OB-16	Sweet basil	Udaipur, Rajasthan (India)	Rich in methyl eugenol
17	OB-17	Zanzibar basil	Tanzania	Highly rich in methyl chavicol
18	OB-18	Indian basil	Bareilly, UP (India)	
19	OB-19	Sweet basil	CSIR-CIMAP, Lucknow UP (India)	
20	OB-20	Indian basil	Lucknow, UP (India)	
21	OB-21	Indian basil	Lakhimpur (Kheri), UP (India)	
22	OB-22	Sweet basil	CSIR-CIMAP, Lucknow UP (India)	
23	OB-23	Sweet basil	Nasik, Maharashtra (India)	
24	OB-24	Sweet basil	Lucknow, UP (India)	
25	OB-25	Indian basil	CSIR-CIMAP, Lucknow UP (India)	
26	OB-26	Sweet basil	CSIR-CIMAP, Lucknow UP (India)	
27	OB-27	Sweet basil	Trivandrum, Kerala (India)	
28	OB-28	Sweet basil	Lucknow, UP (India)	
29	OB-29	Sweet basil	Allahabad, UP (India)	
30	OB-30	French basil	Haridwar, Uttaranchal (India)	
31	OB-31	Sweet basil	CSIR-CIMAP, Lucknow UP (India)	

(continued)

Table 6.2 (continued)

S. No.	Accession No.	Accessions/cultivar	Place of collection/origin	Distinct features
32	OB-32	Thai basil	Thailand	
33	OB-33	Sweet basil	Purulia, WB (India)	
34	OB-34	Sweet basil	CSIR-CIMAP, Lucknow UP (India)	
35	OB-35	Sweet basil	Jammu (J and K) (India)	
36	OB-36	Indian basil	Phagwara, Punjab (India)	
37	OB-37	Sweet basil	Barabanki, UP (India)	
38	OB-38	Sweet basil	Shillong, Meghalaya (India)	
39	OB-39	Indian basil	Razaganj, UP (India)	
40	OB-40	Indian basil	Rishikesh, Uttaranchal (India)	Rich in linalool
41	OB-41	French basil	Bangalore, Karnataka (India)	
42	OB-42	French basil	Mangalore, Karnataka (India)	
43	OB-43	French basil	Chandigarh	
44	OB-44	Sweet basil	CSIR-CIMAP, Lucknow UP (India)	
45	OB-45	Sweet basil	Singapore	
46	OB-46	Sweet basil	Singapore	
47	OB-47	Sweet basil	Singapore	
48	OB-48	Sweet basil	CSIR-CIMAP, Lucknow UP (India)	
49	OB-49	Sweet basil	Košice, Slovak Republic	
50	OB-50	Sweet basil	Gandhi Nagar, Gujarat (India)	High herb yield per plant
51	OB-51	French basil	Bangalore, Karnataka (India)	
52	OB-52	French basil	Mangalore, Karnataka (India)	
53	OB-53	Sweet basil	Košice, Slovak Republic	
54	OB-54	Sweet basil	Gandhi Nagar, Gujarat (India)	High oil content
55	OB-55	Sweet basil	Singapore	
56	OB-56	French basil	Chennai, AP (India)	
57	OB-57	Sweet basil	CSIR-CIMAP, Lucknow UP (India)	
58	OB-58	Sweet basil	Gandhi Nagar, Gujarat (India)	
59	OB-59	Sweet basil	Gandhi Nagar, Gujarat (India)	
60	OB-60	Sweet basil	Singapore	

and methyl eugenol. Thus, these lines can be used when selection is to be made for a line rich in a particular oil component.

CSIR-CIMAP has also released several varieties of *Ocimum* with unique morpho-chemotypic traits.

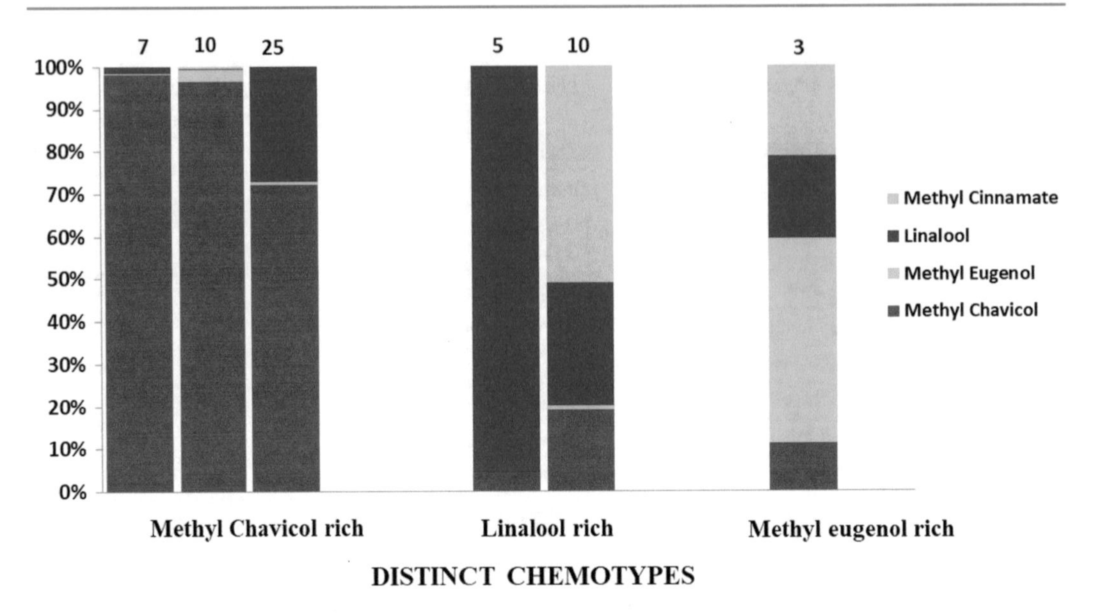

Fig. 6.3 Chemical diversity in *O. basilicum* germplasm accessions at CSIR-CIMAP, Lucknow. Numbers at the top of each bar represent the total number of accessions belonging to that chemotype

Some of the *O. basilicum* varieties released are

- Vikarsudha (developed through intraspecific hybridization between exotic basil from Australia (EC-331886-CSIRO No. L6323) and local adaptive landrace, Badaun local; having 78% methyl chavicol and 0.16% linalool in its oil)
- Kusumohak (developed through selection in seed-raised progeny of the introduced strain from Argentina; having 37% methyl chavicol, 45% linalool in its oil)
- CIMAP-Saumya (developed through half-sib selection, short duration, dwarf, early flowering; having 62.54% methyl chavicol and 24.61 linalool)
- CIMAP-Sharda (developed through intensive breeding, early maturing, high oil yielding, 85–89% methyl chavicol)
- CIMAP-Surabhi (cold tolerant; linalool rich).

Some of the *O. sanctum* varieties released are

- CIM-Kanchan (developed through selection; having 70% methyl eugenol, 7.6% ß-elemene, 15.7% ß-caryophyllene
- CIM-Ayu (developed through mass selection; having 83% eugenol, 7.47% ß-elemene)
- CIM-Angna (developed through half-sib selection; having 40.42% eugenol, 14.11% ß -elemene, 9.07% ß-caryophyllene, 16.65% germacrene-D

Released *O. africanum* variety:

- CIMAP-Jyoti (developed through selection, dwarf; having 68–75% citral).

6.8 Conclusions

The species of the genus *Ocimum* have established its importance both as aromatic and medicinal plant. This genus offers a lot of opportunities on solving the ambiguity regarding the existence of different species and their ploidy level. The molecular markers certainly have played very important role in solving this ambiguity but not beyond doubt. It appears that the species of this genus are still in evolving stage and genome churning is going on giving rise to new chemotypes and probably new natural hybrids/species as well. Considerable genetic variability of this genus has been collected, characterized, and maintained by different *Ocimum* researchers, which might be only a part of the total variability which is still expanding. Several varieties/cultivars belonging to different species of this genus have been developed with specific uses by various researchers exploiting the natural variability. This genus provides an open platform to the plant breeders to exercise their breeder/s skill in developing the genetically improved newer chemotypes of different species for oil quality, oil content, and herb/biomass yield.

References

Aghaei M, Darvishzadeh R, Hassani A (2012) Molecular characterization and similarity relationships among Iranian basil (*O. basilicum*) accessions using inter simple sequence repeat markers. Rev Ciênc Agron 43:312–320

Ahmad SD, Khaliq I (2002) Morpho-molecular variability and heritability in *Ocimum sanctum* genotypes from Northern Himalayan region of Pakistan. Pak J Biol Sci 5(10):1084–1087

Amaral CLF, Casali VWD (2000) Identification and characterization of two populations of 'Alfavaca' (*Ocimum selloi* Benth.) by isozyme markers, Identification and characterization of two populations of 'Alfavaca' (*Ocimum selloi* Benth.) by isozyme markers. Rev Brasil Plantas Med 2(2):9–15

Anbuselvi S, Balamurugan, Kumar S (2013) Purification and characteristics of peroxidase from two varieties of tulsi and neem. Res J Pharma Biol Chem Sci 4 (1):648–654

Balyan SS, Pushpangadan P (1988) A study on the taxonomic status and geographic distribution of the genus, *Ocimum*. Perf Flav Assoc India 10(2):13–19

Bast F, Rani P, Meena D (2014) Chloroplast DNA phylogeography of holy basil (*Ocimum* tenuiflorum) in Indian subcontinent. The Sci World J 2014. http://dx.doi.org/10.1155/2014/847482

Bernhardt B, Szabo K, Bernáth J (2015) Sources of variability in essential oil composition of *Ocimum americanum* and *Ocimum tenuiflorum*. Acta Alimenta 44(1):111–118

Bhamra S, Heinrich M, Howard C, Johnson MRD, Slater A (2014) DNA authentication of Tulsi; the cultural and medicinal value of *Ocimum* species among diasporic South Asian communities in the UK. Planta Med 80(16)

Briquet J (1897) *Ocimum*. In: Engler A, Prantle KAE (eds) Dienatürlichen Pflanzenfamilien, vol 4(3a). Engelmann W, Leipzig, pp 369–372

Carović-Stanko K, Orlic′ S, Politeo O, Strikic′ F, Kolak I, Milos M, Satovic Z (2010) Composition and antibacterial activities of essential oils of seven *Ocimum taxa*. Food Chem 19(1):196–201

Carović-Stanko K, Liber Z, Politeo O, Striki′c F, Kolak I, Milos M, Satovic Z (2011) Molecular and chemical characterization of the most widespread *Ocimum* species. Plant Syst Evol 294:253–262

Chen SY, Dai TX, Chang YT, Wang SS, Ou SL, Chuang WL, Cheng CY, Lin YH, Lin LY, Ku H (2013) Genetic diversity among *Ocimum* species based on ISSR, RAPD and SRAP markers. Aust J Crop Sci 7:1463–1471

Chowdhury T, Mandal A, Roy SC, Sarker DD (2017) Diversity of the genus *Ocimum* (Lamiaceae) through morpho-molecular (RAPD) and chemical (GC–MS) analysis. J Genet Eng Biotechnol 15(1):275–286

Christina VLP, Annamalai A (2014) Nucleotide based validation of *Ocimum* species by evaluating three candidate barcodes of the chloroplast region. Mol Ecol Resour 14:60–68

Darlington CD, Wylie AP (1955) Chromosome atlas of flowering plants. George Allen and Unwin, London

Darrah H (1980) The cultivated basils. Thomas Buckeye, Independence, MO

Datta AK, Mukherjee M, Bhattacharya A, Saha A, Mandal A, Das A (2010) The basils—an overview. J Trop Med Plants 11(2):231–241

De Masi L, Siviero P, Esposito C, Castaldo D, Siano F, Laratta B (2006) Assessment of agronomic chemical and genetic variability in common basil (*Ocimum basilicum* L.). Eur Food Res Technol 223:273–281

De Mattia F, Bruni I, Galimberti A, Cattaneo F, Cassiraghi M, Labra M (2011) A comparative study of different DNA barcoding markers for identification of some members of Lamiaceae. Food Res Intl 44:693–702

de-Wet JMJ (1958) Chromosome number in Plectranthus and related genera. S Afr J Sci 54(6):153–156

Edet OU (2018) Agro-morphological evaluation and analysis of genetic parameters in African basil (*Ocimum gratissimum* L.) in Southeastern Nigeria.In: Crop science society of Nigeria: second national annual conference proceedings, University of Nigeria, Nsukka, Enugu State, Nigeria

Egata DF, Geja W, Mengesha B (2017) Agronomic and bio-chemical variability of Ethiopian Sweet basil (*Ocimum basilicum* L.) accessions. Acad Res J Agri Sci Res 5(7):489–508

Elansary HO, Mahmoud EA (2015) Basil cultivar identification using chemotyping still favored over genotyping using core barcodes and possible resources of antioxidants. J Essen Oil Res 27(1):82–87

Erum S, Naeemullah M, Masood S, Khan MI (2011) Genetic variation in the living repository of *Ocimum* germplasm. Pak J Agric Res 24:42–50

FAO (2013) Genebank Standards for Plant Genetic Resources for Food and Agriculture. Rome

Getsadze GN (1975) A cytological study of the chromosome complex in some species of the genus *Ocimum* L. as initial breeding material. Referativ-nyi Z 3:55–64

Grayer RJ, Kite GC, Goldstone FJ, Bryan SE, Paton A, Putievsky E (1996) Intraspecific taxonomy and essential oil chemotypes in sweet basil, *Ocimum basilicum*. Phytochemistry 43:1033–1039

Harley RM, Brighton CA (1977) Chromosome numbers in the genus *Mentha* L. Bot J Linn Soc 74:71–96

Heath HB (1981) Source book of flavors. AVI, Westport, CT

Ibrahim MM, Aboud KA, Al-Ansary AMF (2013) Genetic Variability among three Sweet Basil (*Ocimum basilicum* L.) varieties as revealed by morphological traits and RAPD markers. World Appl Sci J 24 (11):1411–1419

IPGRI (1993) Diversity for development. International Plant Genetic Resources Institute, Rome, Italy

Khan MS, Bahuguna DK, Kumar R, Kumar N, Ahmad I (2012) Study on genetic variability and heritability in *Ocimum* spp. Hort Flora Res Spect 1(2):168–171

Khosla MK (1989) Chromosomal meiotic behaviour and ploidy nature in genus *Ocimum* (Lamiaceae), "Sanctum group". Cytologia 54:223–229

Khosla MK (1995) Study on inter-relationship, phylogeny and evolutionary tendencies in genus *Ocimum*. Indian J Genet 55(1):71–83

Kumar A, Mishra P, Baskaran K, Shukla AK, Shasany AK, Sundaresan V (2016) Higher efficiency of ISSR markers over plastid psbA-trnH region in resolving taxonomical status of genus *Ocimum* L. Ecol Evol 6(21):7671–7682

Labra M, Miele M, Ledda B, Grassi F, Mazzei M, Sala F (2004) Morphological characterization, essential oil composition and DNA genotyping of *Ocimum basilicum* L. cultivars. Plant Sci 167:725–731

Lawrence BM (1992) Chemical components of Labiatae oils and their exploitation. In: Harley RM, Reynolds T (eds) Advances in labiate sciences. Royal Botanic Gardens, Kew, pp 399–436

Linnaeus C (1753) Species Plantarum Exhibitentes Plantas Rite Cognitas ad Genera Relatas, cum Differentiis Specificis, Nominibus Trivialibus, Synonymis Selectis, et Locis Natalibus, Secundum Systema Sexuale Digestas, edn 1. Laurentius Salvius, Stockholm, Sweden. Facsimile published 1957–1959 as Ray Soc. Publ. 140 and 142. The Ray Society, London, UK

Mahajan V, Rahther IA, Awasthi P, Anand R, Gairola S, Meena SR, Bedi YS, Gandhi SG (2015) Development of chemical and EST-SSR markers for *Ocimum* genus. Indust Crops Prod 63:65–70

Malav P, Pandey A, Bhatt KC, Krishnan SG, Bisht IS (2015) Morphological variability in holy basil (*Ocimum tenuiflorum* L.) from India. Genet Resour Crop Evol 62:1245–1256

Marotti M, Piccaglia R, Giovanelli E (1996) Differences in essential oil composition of basil (*Ocimum basilicum* L.) Italian cultivars related to morphological characteristics. J Agric Food Chem 44:3926–3929

Martins AP, Salgueiro LR, Vila R, Tomi F, Cañigueral S, Casanova J, da Cunha A Proença, Adzet T (1999) Composition of the essential oils of *Ocimum canum, O. gratissimum and O. minimum*. Planta Med 65:187–189

Mehra PN, Gill LS (1972) Cytology of west Himalayan Labiatae, tribe Ocimoideae. Cytologia 37:53–57

Mitra K, Datta N (1967) In: IOPB chromosome number reports XIII. Taxon 16, 163–169

Moghaddam M, Omidbiagi R, Naghavi MR (2011) Evaluation of genetic diversity among Iranian accessions of *Ocimum* spp. using AFLP markers. Biochem Syst Ecol 39:619–626

Morton JK (1962) Cytotaxonomic studies on the West African Labiatae. J Linn Soc Bot 58:231–283

Mukherjee M, Datta AK, Maiti GG (2005) Chromosome number variation in *Ocimum basilicum* L. Cytologia 70(4):455–458

Mukherjee M, Datta AK (2006) Secondary associations in *Ocimum* spp. Cytologia 71(2):149–152

Mukherjee M, Dutta AK (2007) Basils—a review. Plant Arch 7(2):473–483

NBPGR (2002) Minimal descriptors for agri-horticultural crops. Part IV: medicinal and aromatic Plants. NBPGR, New Delhi

Omidbaigi R, Mirzaee M, Hassani ME, Sedghi M (2010) Induction and identification of polyploidy in basil (*Ocimum basilicum* L) medicinal plant by colchicine treatment. Int J Plant Prod 4:87–98

Padmanaban V, Karthikeyan R, Karthikeyan T (2013) Differential expression and genetic diversity analysis using alpha esterase isozyme marker in *Ocimum sanctum* L., Acad. J Plant Sci 6(1):01–12

Patel RP, Kumar RR, Singh R, Singh RR, Rao BRR, Singh VR, Gupta P, Lahri R, Lal RK (2015) Study of genetic variability pattern and their possibility of exploitation in *Ocimum* germplasm. Indust Crops Prod 66:119–122

Paton A (1992) A synopsis of *Ocimum* L (Labiatae) in Africa. Kew Bull 47:405–437

Paton A, Putievsky E (1996) Taxonomic problems and cytotaxonomic relationships between and within varieties of *Ocimum basilicum* and related species (Labiatae). Kew Bull 51:509–524

Paton A, Harley MR, Harley MM (1999) *Ocimum*: an overview of classification and relationships. In: Hiltunen R, Holm Y (eds) Basil: the genus *Ocimum*. Harwood, Amsterdam, pp 1–38

Puri HS (2002) Rasayana: Ayurvedic herbs for longevity and rejuvenation. CRC Press, Boca Raton, pp 272–280

Pushpangadan P (1974) Studies on reproduction and hybridisation of *Ocimum species* with view to improving their quality. Ph.D. thesis, Aligarh Muslim University, Aligarh, India

Putievsky E, Paton A, Lewinsohn E, Ravid U, Haimovich D, Katzir I, Saadi D, Dudai N (1999) Crossability and relationship between morphological and chemical varieties of *Ocimum basilicum* L. J Herbs Species Med Plants 6:11–24

Pyne R, Honig J, Vaiciunas J, Koroch A, Wyenandt C, Bonos S et al (2017) A first linkage map and downy mildew resistance QTL discovery for sweet basil

(*Ocimum basilicum*) facilitated by double digestion restriction site associated DNA sequencing (ddRAD-seq). PLoS ONE 12(9):e0184319. https://doi.org/10.1371/journal.pone.0184319

Ravid U, Putievsky E, Katzir I, Lewinsohn E (1997) Enantiomeric composition of linalol in the essential oils of *Ocimum* species and in commercial basil oils. Flav Fragr J 12:293–296

Rewers M, Jędrzejczyk I (2016) Genetic characterization of *Ocimum* genus using flow cytometry and inter-simple sequence repeat markers. Indust Crops Prod 91:142–151

Sadeghi S, Rahnavard A, Ashrafi ZY (2009) The effect of plant density and sowing date on yield of basil (*Ocimum basilicum* L.) in Iran. J Agric Technol 5(2):413–422

Shinde K, Shinde V, Mahadik K, Gibbons S (2010) Phytochemical and antibacterial studies on *Ocimum kilimandscharicum*. Planta Med 76(Suppl 1):1295

Simon JE, Quinn J, Murray RG (1990) Basil: a source of essential oils. In: Janick J, Simon JE (eds) Advances in new crops. Timber, Portland, pp 484–989

Singh S, Lal R K, Maurya R, Chanotiya CC (2018) Genetic diversity and chemotype selection in genus *Ocimum*. J Appl Res Med Arom Plants. https://doi.org/10.1016/j.jarmap.2017.11.004

Singh TP (1978) Chromosome studies in *Ocimum*. CurR Sci 47(23):915–916

Singh YP, Gaurav SS, Kumar P, Ojha A (2013) Genetic variability, heritability and genetic advance in segregating generations of Basil (*Ocimum basilicum* L.). Progr Res 8(2):164–168

Sobti S, Pushpangadan P (1977) Cytotaxonomical studies in the genus I. In: Bir SS (ed) Taxonomy, cytogenetics, cytotaxonomy of plants. Kalyani Publisher, New Delhi, India, pp 373–377

Sobti SN, Pushpangadan P, Atal CK (1982) Clocimum: A new hybrid strain of *Ocimum gratissimum* as a potential source of clove type oil, rich in eugenol. In: Atal CK, Kapur BM (eds) Cultivation and utilization of aromatic plants. Regional Research Laboratory, Jammu-Tawi, India, pp 473–480

Srivastava A, Gupta AK, Sarkar S, Lal RK, Yadav A, Gupta P, Chanotiya CS (2018) Genetic and chemotypic variability in basil (*Ocimum basilicum* L.) germplasm towards future exploitation. Indust Crops Prod 112:815–820

Upadhyay AK, Chacko AR, Gandhimathi A, Ghosh P, Harini K et al (2015) Genome sequencing of herb Tulsi (*Ocimum tenuiflorum*) unravels key genes behind its strong medicinal properties. BMC Plant Biol 15:212

Upadhyaya HD, Gowda CLL, Sastry DVSSR (2008) Plant genetic resources management: collection, characterization, conservation and utilization. J SAT Agric Res 6

UPOV (2003) Basil (*Ocimum basilicum* L.): Guidelines for the conduct of tests for distinctness, uniformity and stability. Geneva

Vij SP, Kashyap SK (1976) Cytological studies in some North Indian Labiatae. Cytologia 41(3–4):713–719

Wealth of India (1976) A dictionary of Indian raw materials and industrial products, vol III. Publications and Informations Directorate, CSIR, New Delhi, pp 77–89

Werker E, Putievsky E, Ravid U, Dudai N, Katzir I (1993) Glandular hairs and essential oil in developing leaves of *Ocimum basilicum* L. (Lamiaceae). Ann Bot 71(1):43–50

Wetzel SB, Krüger H, Hammer K, Bachmann K (2008) Investigations on morphological, biochemical and molecular variability of *Ocimum* L. Species. J Herbs Spices Med Plants 9(2–3):183–187

Traditional Plant Breeding in *Ocimum*

7

R. K. Lal, Pankhuri Gupta, C. S. Chanotiya and Sougata Sarkar

Abstract

Ocimum spp. is an important medicinal and aromatic plant. It has many medicinal properties. It is a rich source of carbohydrates, fibers, iron, β-carotene, vitamins, phosphorous, calcium, protein, and in aromatic oils. It is also used for the treatment in stomach pain, cough and cold, diarrhea and indigestion. Asthma, ulcers, nausea, and ringworm can also be cured with *Ocimum*. It lowers the blood sugar level and increases lactation. CSIR-CIMAP is actively involved in genetic enhancement of the *Ocimum* species following with different breeding approach in view of traditional importance. At CSIR-CIMAP, available genetic stocks are seven *Ocimum* species—*Ocimum sanctum*—Krishna and Shyam tulsi, *O. basilicum*, *O. kilimandscharicum*, *O. americanum*, *O. africanum*, *O. gratissimum*, *O. tenuiflorum,* and 100 genetic stocks of *O. basilicum* and nine varieties, namely CIM Ayu, CIM Angana, CIM Saumya, CIM Kanchan, Vikarsudha, CIM Jyoti, CIM Sharada, CIM Surabhi, and CIM Snigdha. In future, there will be possibility to develop varieties for high oil and herb yield with high specific needs chemical like high eugenol, methyl eugenol, methyl cinnamate, geraniol, germacrene A and D, linalool, elemicin, ß-elmene, (Z)-ocimine content with some other herbal products.

R. K. Lal (✉)
Genetics and Plant Breeding Division, CSIR-Central Institute of Medicinal and Aromatic Plants, Lucknow 226015, India
e-mail: rajkishorilal@gmail.com

P. Gupta
Biotechnology Division, CSIR-Central Institute of Medicinal and Aromatic Plants, Lucknow 226015, India

C. S. Chanotiya
Analytical Chemistry Division, CSIR-Central Institute of Medicinal and Aromatic Plants, Lucknow 226015, India

S. Sarkar
Genetic Resources and Agro-tech Division, CSIR-Indian Institute of Integrative Medicine, Jammu 180001, India

7.1 Introduction

Ocimum belongs to Lamiaceae family, and it is a genus of about 180–250 species of annual and perennial aromatic herbs and shrubs. Several species are native to the tropical and warm temperate regions of the old world, including India. The dry herb (leaves), *Ocimum* leaf tea, essential oil and its chemical derivatives (eugenol, methyl–eugenol, linalool, methyl chavicol, germacrene A and D, elemicin, ß-elmene, (Z)-ocimine) are exported to European countries in extensive quantity every year (Simon et al. 1990). The

A. K. Shasany and C. Kole (eds.), *The Ocimum Genome*, Compendium of Plant Genomes, https://doi.org/10.1007/978-3-319-97430-9_7

Fig. 7.1 Released varieties/cultivars from CSIR-CIMAP of *Ocimum*

annual export of dry leaves herb, its products, essential oil, and its derivatives/chemical constituents of *Ocimum* are worth 5000 tons. People know the plant as Tulsi as *Surasah* in Sanskrit and *Tulsi* in Hindi. Due to antioxidant and anti-aging properties of Tulsi, Hindus use fresh leaves in the *Panchamrut/Charanamrut* drink after *puja*. Tulsi is religion. It is regarded not merely as a utilitarian Godsend, as most holy plants are viewed to be, but as an incarnation of the Goddess Herself. The classic Hindu myth, *Samudramathana*, the 'Churning of the Cosmic Ocean,' explains that *Vishnu* spawned Tulsi from the turbulent seas as a vital aid for all mankind. The Tulsi leaves, when eaten, can control thirst, and so was invaluable to tired travelers. The oil is used as antiperspirant and as fly and mosquito repellent. CSIR-CIMAP is actively involved in genetic enhancement of the *Ocimum* species by using different breeding approaches in view of traditional significance, together with the need for developing a better plant type having high herb, essential yield characters combined with a consistent high yield of phenylpropanoid eugenol and other economically important chemical constituents to formulate value-added products. By the application of very intensive plant breeding techniques, a number of varieties have been developed at the CSIR-Central Institute of Medicinal and Aromatic Plants, India, including: Khushmohak, CIM Saumya (Lal et al. 2003), CIM Sharda of *Ocimum basilicum*, CIM Jyoti of *O. africanum*, CIM Ayu (Lal et al. 2004), CIMAngana of *Ocimum sanctum* (Lal et al. 2008), CIM Kanchan from *O. tenuiflorum*, CIM Surabhi, and CIM Snigdha (Fig. 7.1). These varieties/ cultivars are also useful for intercropping with aromatic grasses and other cereal crops. Some of the international popular varieties/ cultivars and released varieties/cultivars from CSIR-CIMAP of *Ocimum* are summarized in Tables 7.1 and 7.2.

By conventional breeding method, Lal et al. (2003) developed a high-yielding eugenol-rich oil producing variety of *O. sanctum*, CIM Ayu; Lal et al. in 2004 developed an early, short duration, high essential oil, methyl chavicol and linalool yielding variety of Indian basil (*O. basilicum*), CIM Saumya; Lal et al. (2008) also developed a high-yielding dark purple pigmented variety CIM Angana of Shyam tulsi (*Ocimum sanctum* L.); Singh and Sehgal (1999) chemically characterized and made selection of *Ocimum* genotypes and described their uses in traditional medicine. They studied the genetic variability, medicinal, and economic value of *Ocimum* germplasm. Lal (2014) developed new and stable chemotypes in *Ocimum*. Patel et al. (2015a, b) studied genetic variability pattern and possibility of its exploitation of *Ocimum* germplasm in Fig. 7.2. Patel et al. (2015a, b) also studied the phenotypic characterization and

Table 7.1 International popular varieties/cultivars of *Ocimum* species

S. no.	Varieties/cultivars	Botanical identification	Origin of the plant material	Chemotypes	Breeding strategies
1.	Blue Spice	*Ocimum basilicum*	Czech Republic	Bisabolene	Asexual reproduction
2.	Fino Verde	*Ocimum basilicum*	Maine, USA	Linalool	Asexual reproduction
3.	Holandjanin	*Ocimum basilicum*	Istria	Linalool	Asexual reproduction
4.	Compact	*Ocimum basilicum*	Maine, USA	Linalool	Asexual reproduction
5.	Genovese	*Ocimum basilicum*	Slovenia	Linalool	Selection
6.	Purple Opal	*Ocimum basilicum*	Czech Republic	Linalool	Selection
7.	Lattuga	*Ocimum basilicum*	USA	Linalool	Selection
8.	Osmin	*Ocimum basilicum*	California	Linalool	Selection
9.	Cinnamon	*Ocimum basilicum*	Wroclaw, Poland	Linalool and methyl cinnamate	Selection
10.	Purple Ruffles	*Ocimum basilicum*	Germany	Linalool	Cross selection
11.	Lime	*Ocimum americanum*	Czech Republic	Geranial and neral	Selection
12,	Siam queen	*Ocimum basilicum*	USA	Methyl chavicol	Cross selection
13.	Dark Lady	*Ocimum basilicum*	Wolsier, Germany	Linalool	Asexual reproduction
14.	Dbasbloom	*Ocimum basilicum*	Israel	Linalool	Asexual reproduction
15.	Green Bell	*Ocimum basilicum*	Wolsier, Germany	Linalool	Asexual reproduction

stability analysis for biomass and essential oil yields of 15 genotypes of five *Ocimum* species. Patel et al. (2016) studied the differential response of genotype × environment on phenology, essential oil yield, and quality of natural aroma chemicals of five *Ocimum* species. Srivastava et al. (2018) conducted germplasm characterization and correlation studies in *O. basilicum* for yield and related traits. Singh et al. (2004, 2014) chemically characterized and made selection in *Ocimum* genotypes.

7.2 Breeding for Disease Resistance

Breeding work for disease resistance in *Ocimum* is very meager as described below.

Mishra et al. (2016) reported the identification and functional characterization of an *O. basilicum* PR5 family member that is responsive to multiple stresses and hormonal elicitations. The results highlighted the role of a thaumatin-like protein, ObTLP1, in mediating tolerance to the

Table 7.2 Released varieties/cultivars from CSIR-CIMAP of *Ocimum* species with their quality

S. no	Varieties/cultivars	Botanical identification	Common name	Herb yield (ql/ha)	Oil content (%)	Oil yield (kg/ha)	Main essential oil constituents						Breeding strategy
							Eugenol	Methyl eugenol	Methyl chavicol	Linalool	Methyl cinnamate	Citral content	
1.	Vikarsudha	*Ocimum basilicum*	French basil	335	0.5	167.5	0.62	–	78	0.16	–	–	Developed through intraspecific hybridization between exotic basil from Australia (EC-331886-CSIRO No. L6323) and local adaptive landrace, Badaun local
2.	Khushmohak	*Ocimum basilicum*	Sweet basil	391	0.4	134	–	–	37	45	–	–	Developed through selection in seed raised progeny of the introduced strain from Argentina
3.	CIM Ayu	*Ocimum sanctum*	Holy basil (Krishna)	200	0.72	110.95	83.56	–	–	0.05	–	–	Developed through mass selection
4.	CIM Angana	*Ocimum sanctum*	Holy basil (Shyama)	181	0.64	91.7	40.42	–	–	1.92	–	–	Developed through half-sib selection
5.	CIM Kanchan	*Ocimum teniflorum*	Holy basil	197	0.49	110.7	4.36	78.4	–	0.09	–	–	Developed through selection
6.	CIM Saumya	*Ocimum basilicum*	Indian basil	290	0.0.68	197.2	–	–	62.54	24.61	–	–	Developed through half-sib selection
7.	CIM Jyoti	*Ocimum africanum*	Lemon-scented basil	200	0.75	150	–	–	–	–	–	75	Developed through selection
8.	CIM Sharada	*Ocimum basilicum*	French basil	280	0.7	190	–	–	89.75	0.067	–	–	Developed through intensive breeding
9.	CIM Shurabhi	*Ocimum basilicum*	Sweet basil	200	0.75	166	–	–	0.44	75.71	–	–	Developed through half-sib selection
10.	CIM Snigdha	*Ocimum basilicum*	French basil	221	0.9	190	–	–	9.1	1.95	78.7	–	Developed through half-sib selection

Fig. 7.2 Variability among leaves of the genetic stocks of *Ocimum*

fungal pathogens and to the abiotic stresses by hindering fungal colonization in host and by maintaining osmotic adjustment in cells, respectively, and ObTLP1 might be a useful candidate for providing tolerance in crops toward multiple stresses.

Zaim et al. (2016) studied the occurrence of yellow mosaic and leaf curl disease in *Ocimum* spp. and conducted screening for disease management. They investigated that yellow mosaic and leaf curl diseases affected *O. killimandscharium*, *O. gratissimum* and very severely to *O. basilicum* var. CIM Saumya in non-homozygous population. Investigation has also been carried out to evaluate the available germplasm of *Ocimum* spp. for identifying resistance source. Since *O. basilicum* was found to be highly susceptible to these viral diseases, a few lines have been selected, which were found to be tolerant against the leaf curl disease. These studies will help in strategizing the effective management practices and developing resistant cultivars in *Ocimum*.

7.3 Polyploidy in Basil

Sobti and Pushpangadan (1982) presented in detail the cytogenetical and evolutionary relationship in *Ocimum americanum*, *O. canum,* and *O. basilicum. O. americanum* was found to be morphologically intermediate between the other two species in most characters like height, length of spike, calyx, corolla, stamen, style fruiting calyx. Crossability studies and induction of alloploidy were also attempted by above workers. Karyomorphological investigations were done by Sobti and Pushpangadan (1982), in five *Ocimum* species with their different races. Closer examination of the karyotypes revealed variation in the type and number of chromosomes and considered the presumption as mentioned in Chap. 6. Meiotic studies in five species of *Ocimum* from pollen mother cells by Khosla (1989) suggested that the base number $x = 8$ represented Sanctum group while base number $x = 12$ represented Basilicum group. Omidbaigi et al. (2010) studied the induction and identification of polyploidy in *O. basilicum* using colchicines in different concentrations (0.00, 0.05, 0.10, 0.20, 0.50, and 0.75%) and four treatments (seed, the growing point of seedlings at the emergence of cotyledon leaves stage and emergence of true two type leaves stage, and root treatment). The 0.5% dosage proved to be the most effective in producing autotetraploids having larger stomata and pollen grains, increased chloroplast number in guard cells, and decreased stomata density, compared to diploid control plants.

7.4 Linalool Rich, Evergreen, and Cold-Tolerant Interspecific Cross Hybrid

A cold-tolerant evergreen triploid and its amphidiploid also developed using interspecific cross between *O. basilicum* and *O. kilimandscharicum* are under field evaluation trial for their essential oil yield with good quality (Fig. 7.3). It is an evergreen aromatic perennial undershrub. Improving the cold tolerance and developing resistant varieties of *Ocimum* are very important for the growing areas of the species (Dhawan et al. 2015). In relation to adaptive behavior, *O. basilicum* and other species are sensitive to cold stress except *O. kilimandscharicum*; therefore, an evergreen interspecific hybrid was developed by hybridization through breeding efforts from *O. basilicum* (CIM Surabhi) with *O. kilimandscharicum* for stress tolerance (Fig. 7.4). The developed evergreen interspecific hybrid may thus provide a base to various industries which are dependent upon the bioactive constituents of *Ocimum* species, and one another cold-tolerant evergreen interspecific cross between *O. basilicum* cultivar CIM Surabhi and *O. kilimandscharicum* (−) linalool 75–80% is also under pipeline and under field evaluation trial for their essential oil yield with good quality (−) linalool.

7.5 A Novel Source of Chavibetol Constituent in *Ocimum basilicum*

CSIR-CIMAP, Lucknow, has also developed a new line with chavibetol (8–10%) proportions and flavor of *Ocimum* leaves as paan/betel (*Piper betle*) (Fig. 7.6). Eugenol was recorded in this accession in trace amount. Structure of chavibetol was confirmed using NMR experiments (Fig. 7.5).

7.6 Basil as an Edible Ornamental Herb

7.6.1 Commercial Cultivars

Several basil varieties, differing in the size, shape, aroma, and color of the leaves, exist. Commercial basil cultivars also display a wide diversity in growth habit, flower, leaf, and stem colors, and aroma. Many of the cultivars evaluated belong to the ‘sweet’ basil group with ‘Genovese,’ ‘Italian large leaf,’ ‘Mammoth,’ ‘Napoletano,’ and ‘Sweet’ dominating the American fresh and dry culinary herb markets. Several other basils like ‘Sweet Fine’ appear similar to ‘sweet’ basil

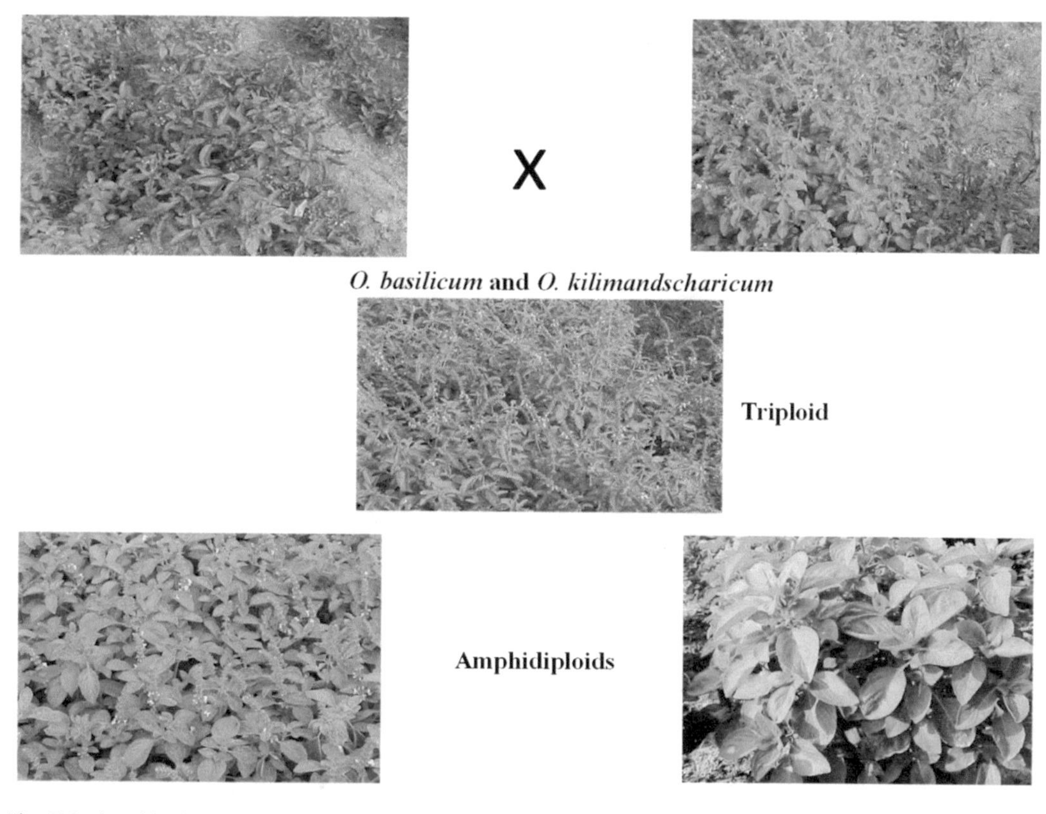

Fig. 7.3 A cold-tolerant evergreen triploid and its amphidiploid also developed using interspecific cross between *O. basilicum* and *O. kilimandscharicum* are under field evaluation trial for their essential oil yield with good quality

Fig. 7.4 A (−) linalool-rich cold-tolerant evergreen interspecific cross between *O. basilicum* cultivar CIM Surabhi and *O. kilimandscharicum* (−) linalool 75–80%

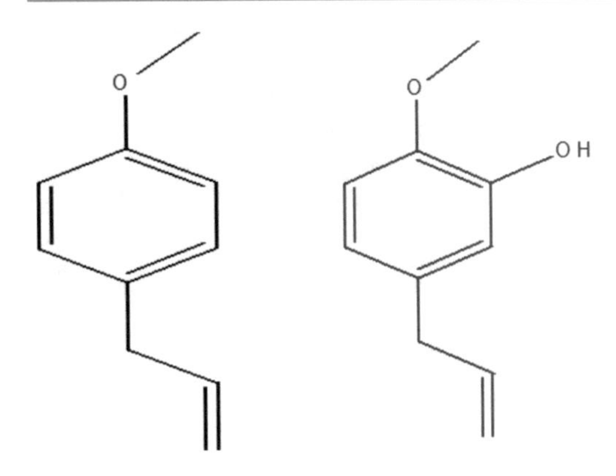

Fig. 7.5 Structure of chavibetol

Fig. 7.6 OCL-32 chavibetol in *Ocimum*

though the leaves tend to be smaller (Simon et al. 1999). The lemon-scented cultivars ('Lemon' and 'Lemon Mrs. Burns') differ from each other in days to flower, and total oil content, but not in citral content. The 'Maenglak Thai Lemon' basil, which varies in appearance from the other lemon basils, is an attractive ornament. Among the purple basils, 'Osmin Purple' and 'Red Rubin Purple Leaf' are the most attractive and best retain their purple leaf color (Simon et al. 1999). Anthocyanins in purple basils are genetically unstable, leading to an undesirable random green sectoring and reversion over the growing season (Phippen and Simon 1998). Several basils with dwarf growth habit were developed as ornamental border plants including 'Bush,' 'Green Globe,' 'Dwarf Bush,' 'Spicy Globe,' and 'Purple Bush.' A group of ornamental basils was selected and named for their characteristic aroma including 'Anise' (methyl chavical), 'Cinnamon' (methyl cinnamate), 'Licorice' (methyl chavicol), and 'Spice' (bisabolene) (Darrah 1972; Albuguerque 1996; Simon et al. 1999).

7.6.2 Sensoric Quality

Fresh basil leaves have a strong and characteristic aroma, not comparable to any other spice, although a traceable hint of cloves exists. In addition to the 'Mediterranean type' most common in the West, a plethora of other varieties or cultivars with different flavors, many of which are hybrids, is available. India has 'sacred basil' (*O. sanctum* = *O. tenuifolium*) with an intensive, somewhat pungent smell, and Thailand has a sweet basil with a licorice aroma (Singh and Sehgal 1999). Varieties sold to gardeners in the West include cinnamon basil, camphor basil, anise basil, and spice basil; the latter has a very pleasant, complex, and warm flavor (Darrah 1974; Albuguerque 1996; Morales and Simon 1996; Phippen and Simon 1998; Martins et al. 1999; Simon et al. 1999). A last group of cultivars, characterized by citrus aroma, are 'Thai Lemon basil' (*O. citriodorum*) which has a distinct balm-like flavor and lime basil and another lemon basil (*O. americanum*) which has an extraordinarily pure and fresh lemon aroma (Morales and Simon 1996). Perennial basil species from Africa (*O. kilimandscharicum*) and Asia (*O. canum*) have recently been introduced to the European herb and gardening market. These species have a strong, but less pleasant flavor, and hybrids between them and Mediterranean basil are a recent innovation with a novel appearance and flavor that are enjoying a growing popularity (Darrah 1972; Kanebo 1992; Grieve 1999). All basil varieties have in common that dried leaves are much less aromatic than fresh ones; deep freezing the herb is the best method of preservation.

7.6.3 Ecology

Sweet basil is cultivated in agro climes between 7 and 27 °C, with 0.6–4.2 m annual precipitation and soil pH 4.3–8.2. While susceptible to frost and cold temperature injury, the species develop best under long days, in full sun and on well-drained soils.

7.7 Future Prospects

The conventional function of plant breeding in the area of feed, food, fiber, and ornate will persist to be significant. In spite of this, new functions are progressively budding for plants. The skill of using plants as bioreactors for generating pharmaceuticals will expand; this knowledge is therefore over a decade. Approaches for using plants for generating medicinal antibodies, engineering antibody-mediated pathogen resistance, and altering plant phenotypes by immunomodulation are being refined. New skills are being expounded for plant breeders, particularly, in the areas of the biotechnological applications to plant breeding. New marker knowledge is going on to be refined, and older ones are progressive. Tools assisting the plant breeders to be more efficiently controlling the quantitative traits will be improved.

References

Albuguerque U (1996) Taxonomy and ethnobotany of the genus *Ocimum*. Federal University of Pernambuco, Mexico, pp 48–68

Dhawan SS, Shukla P, Gupta P, Lal RK (2015) A cold-tolerant evergreen interspecific hybrid of *Ocimum kilimandscharicum* and *Ocimum basilicum*: analyzing trichomes and molecular variations. Protoplasma 253:845–855

Darrah HH (1972) The basils in folklore and biological science. Herbalist 38:3–10

Darrah HH (1974) Investigations of the cultivars of basils (*Ocimum*). Econ Bot 28:63–67

Grieve M (1999) A modern herbal. Tiger Books International, London, pp 85–87

Kanebo C (1992) Basil propagation from seedling. Patent application Nr. JP006013

Khosla MK (1989) Chromosomal meiotic behaviour and ploidy nature in genus *Ocimum* (Lamiaceae), "Sanctum group". Cytologia 54:223–229

Lal RK (2014) Breeding for new chemotypes with stable high essential oil yield in *Ocimum*. Indust Crops Prod 59:41–49

Lal RK, Khanuja SPS, Agnihotri AK, Misra HO, Shasany AK, Naqvi AA, Dhawan OP, Kalra A, Bahl JR, Darokar MP (2003) High yielding eugenol rich oil producing variety of *Ocimum sanctum*-'CIM-Ayu'. J Med Arom Plant Sci 25:746–747

Lal RK, Khanuja SPS, Agnihotri AK, Shasany AK, Naqvi AA, Dwivedi S, Misra HO, Dhawan OP, Kalara A, Singh A, Bahl JR, Singh S, Patra DD, Agarwal S, Darokar MP, Gupta ML, Chandra R (2004) An early, short duration, high essential oil, methyl chavicol, and linalool yielding variety of Indian Basil (*Ocimum basilcum*) 'CIM- Saumya'. J Med Arom Plant Sci 26:77–78

Lal RK, Khanuja SPS, Rizavi H, Shasany AK, Ahmad R, Chandra R, Naqvi AA, Misra HO, Singh A, Singh N, Lohia RS, Bansal K, Darokar MP, Gupta AK, Kalara A, Dhawan OP, Bahl JR, Singh AK, Shankar H, Kumar D, Alam M (2008) Registration of a high yielding dark purple pigmented, variety 'CIM-Angana' of Shyam tulsi (*Ocimum sanctum* L.). J Med Arom Plant Sci 30(1):92–94

Martins AP, Salgueiro LR, Vila R, Tomi F, Canigueral S, Casanova J, Proenca da Cunha A, Adzet T (1999) Composition of the essential oils of *O. canum, O. gratissimum* and *O. minimum*. Planta Med 65:187–189

Misra RC, Sandeep Kamthan M, Kumar S, Ghosh S (2016) A thaumatin-like protein of *Ocimum basilicum* confers tolerance to fungal pathogen and abiotic stress in transgenic Arabidopsis. Sci Rep 6:25340

Morales MR, Simon JE (1996) New basil selections with compact inflorescence for the ornamental market. In: Janick J (ed) Progress in new crops. ASHS Press, Alexandria, VA, pp 543–546

Omidbaigi R, Mirzaee M, Hassani ME, Sedghi M (2010) Induction and identification of polyploidy in basil (*Ocimum basilicum* L.) medicinal plant by colchicine treatment. Int J Plant Prod 4:87–98

Patel RP, Kumar RR, Singh R, Singh RR, Rao B, Singh VR, Gupta P, Lahiri R, Lal RK (2015a) Study of genetic variability pattern and their possibility of exploitation in *Ocimum* germplasm. Indust Crop Prod 66:119–122

Patel RP, Singh R, Saikia SK, Sastry KP, Rao BRR, Zaim M, Lal RK (2015b) Phenotypic characterization and stability analysis for biomass and essential oil yields of fifteen genotypes of five *Ocimum* species. Indust Crop Prod 77:21–29

Patel RP, Singh R, Rajeswara BRR, Singh RR, Srivastava A, Lal RK (2016) Differential response of genotype × environment on phenology, essential oil yield and quality of natural aroma chemicals of five *Ocimum* species. Indust Crop Prod 87:210–217

Phippen WB, Simon JE (1998) Anthocyanins in basil. J Agric Food Chem 46:1734–1738
Simon JE, Quinn J, Murray RG (1990) Basil, a source of essential oils. In: Janick J, Simon JE (eds) Advances in new crops. Timber Press, Portland, pp 484–489
Simon JE, Morales MR, Phippen WB, Vieira RF, Hao Z (1999) Basil, a source of aroma compounds and a popular culinary and ornamental herb. In: Janick J (ed) Perspectives on new crops and new uses. ASHS Press, Alexandria, VA, pp 499–505
Singh NK, Sehgal CB (1999) Micropropagation of 'Holy Basil' (*Ocimum sanctum* Linn.) from young inflorescences of mature plants. Plant Growth Regul 29:161–166
Singh S, Singh A, Singh UB, Patra DD, Khanuja SPS (2004) Intercropping of Indian basil (*Ocimum basilicum* L.) for enhancing resource utilization efficiency of aromatic grasses. J Spices Arom Crops 2:97–101
Singh S, Sarkar S, Lahiri R, Chandra R, Lal RK (2014) Study of genetic variability, medicinal and economic value of *Ocimum* germplasm. In: National conference and exhibition on devine medicinal plants, plants based medicinal products, Panchagevya and their role in integrated rural development 21–22 Feb, 2014 at Vigyan Bhawan, New Delhi. Organized by Council for Development of Rural Areas 2 Vigyan Lok, Delhi-92, India Co-Organizer CSIR, New Delhi, CSIR-NBRI, CSIR-CIMAP, Lucknow. Souvenir, pp 35–36
Sobti SN, Pushpangadan P (1982) Studies in the genus *Ocimum*, cytogenetics, breeding and production of new strains of economic importance. In: Atal CK, Kapur BM (eds) Cultivation and utilization of aromatic plants. Regional Research, Laboratory, Council of Scientific and Industrial Research, Jammu-Tawi, India, pp 457–472
Srivastava A, Gupta AK, Sarkara S, Lal RK, Yadav A, Gupta P, Chanotiya CS (2018) Genetic and chemotypic variability in basil (*Ocimum basilicum* L.) germplasm towards future exploitation. Indust Crop Prod 112:815–820
Zaim M, Pandey R, Singh R, Smita Singh, Anju Yadav, Patel RP, Lal RK (2016) Occurrence of yellow mosaic and leaf curl disease of *Ocimum* sp. and screening for disease management. Indian Phytopathol 69: 415–418

Genomic Resources of *Ocimum*

8

Saumya Shah, Shubhra Rastogi and Ajit Kumar Shasany

Abstract

The genus *Ocimum* belonging to the family Lamiaceae, collectively known as basil, has long been acclaimed for the genetic diversity of the species within the genus. *Ocimum* comprises at least 65 species but more than 150 species, according to some sources, of herbs and shrubs from the tropical and subtropical regions of Asia, Africa, and Central and South America. The main center of diversity appears to be Africa. Basil, one of the most popular herbs grown in the world, is native to Asia and can be observed growing wild in tropical and subtropical regions. In the present chapter, an effort has been made to list all the available genomic resources of the *Ocimum* species worldwide and to describe the medicinal potentialities, uses, and essential oil components of some important ones.

8.1 Introduction

Basil is the common name which represents the whole *Ocimum* genus belonging to the mint family, Lamiaceae (formerly known as Labiatae). The basil leaf is a part of sacred traditions worldwide, starting from Hinduism to Christianity. Even though there is no discussion of basil in the holy *Bible* (Darrah 1980), the basil plant is found to be grown at the location of the crucifixion of Jesus Christ (Meyers 2003). *Ocimum tenuiflorum* (Holy basil) is predominantly revered in Indian Hindu custom. It symbolizes the Goddess, Tulasi, and is thought to be originated from her embers. There are several stories regarding the origin of tulsi in the mythology, but as per a common one, Tulasi was mislead to betray her husband when the lord Vishnu in disguise of her husband seduced her. In immense pain, she committed suicide, and in the response lord Vishnu affirmed that 'all women would worship her for her faithfulness' and her blessings would prevent women from becoming widows (Gupta 1971). Therefore, holy basil, popular as tulsi and actually referred to as a goddess, has gradually turned out to be an emblem of eternity, love, purity, and protection among Hindus (Gupta 1971; Thiselton-Dyer 1889). Besides other basil roles, it also has importance in burial rituals; thus, holy basil is found to be grown on or near the graves in different countries (Darrah 1980).

Ocimum is the member of the subfamily Nepetoideae that accommodates plants which are fervently aromatic due to the presence of essential oils containing monoterpenes, sesquiterpenes, and phenylpropenes. The majority of the

S. Shah · S. Rastogi · A. K. Shasany (✉)
Plant Biotechnology Division, CSIR-Central Institute of Medicinal and Aromatic Plants, Lucknow 226015, India
e-mail: ak.shasany@cimap.res.in

A. K. Shasany and C. Kole (eds.), *The Ocimum Genome*, Compendium of Plant Genomes, https://doi.org/10.1007/978-3-319-97430-9_8

basils are rich in phenylpropanoids and the main constituents being eugenol, methyl eugenol, chavicol, methyl chavicol, and linalool. The quantity of each of the constituents differs with different species or varieties. Methyl chavicol imparts a sweet flavor comparable to anise and also French tarragon; linalool gives off a floral aroma; and eugenol is redolent of cloves. Plants of the genus have quadrangular stems, odorous opposite leaves, and flower whorls arranged on the spike of its verticillaster inflorescence (Darrah 1980). The presence of specialized glands known as glandular trichomes on the leaf surface is one of the marked traits of family Lamiaceae, of which the basil is also a member. Two classes of glandular trichomes can be found in these plants (Gang et al. 2002). These are the globular peltate and thread-like capitate glandular trichomes which could be differentiated by their size and number of head cells present. Phenylpropenes in the essential oil are exclusively produced in the peltate glands (Deschamps and Simon 2010).

Even though basil is grown in different agro-climatic conditions, the warm climate is the most suitable for it. Appropriate light, moisture, and warmth are the minimal requirements for cultivating basil. Formerly grown in the Middle East and Asia, basil has passed all around the world along the spices trail. Basil has been grown and used nearly past 5000 years. Having numerous varieties, it is presently being grown in different countries. All the basil varieties have exclusive and unique chemical composition; yet the primary healing properties remain constant. Basil is documented for its endurance against inconsistent ecological conditions and thus is naturalized almost all over the world.

8.2 Status of *Ocimum* Species Worldwide

The status of *Ocimum* species worldwide was examined using '*The Plant List*,' as it provides a working list of every identified plant species and intends to be widespread for vascular plant species (flowering plants, ferns, conifers, and allied ones) as well as of the bryophytes (liverworts and mosses). The Plant List makes available the Latin names for most of the species which are accepted; additionally, it provides links to all possible synonyms by which that particular species has been identified (Fig. 8.1). It also maintains account of uncertain names for which the concerning data source did not hold satisfactory evidence to determine whether they were synonyms or accepted.

The Plant List comprises 333 species rank scientific plant names for the genus *Ocimum.* Among these species names that tend to be accepted are only 68. Moreover, 44 infraspecific rank plant scientific names for the genus *Ocimum* are mainly included as the species rank names were synonymous to accepted infraspecific names (Table 1).

The *Ocimum* species accepted worldwide with their geographical distribution (http://en.wikipedia.org/wiki/Ocimum) are listed as follows:

1. *O. waterbergense*—South African Northern Province
2. *O. viphyense*—Zambia, Malawi

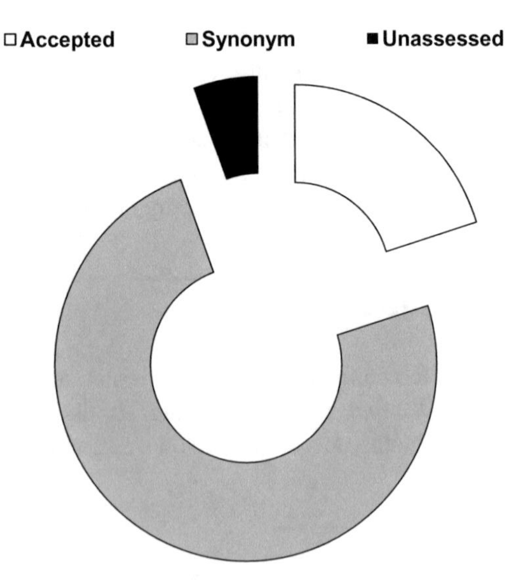

Fig. 8.1 Status of the genus *Ocimum* in *The Plant List* database (http://www.theplantlist.org/1.1/browse/A/Lamiaceae/Ocimum/). Out of these names, 12 are stated as invalid and 16 as illegitimate

Table 8.1 Level of confidence showing the status of names of *Ocimum* species in *The Plant List*

Level of confidence	Accepted	Synonym	Unplaced	Unassessed	Total
High	68	235	0	0	303 (91.0%)
Medium	0	4	0	0	4 (1.2%)
Low	0	2	18	6	26 (7.8%)

http://www.theplantlist.org/1.1/browse/A/Lamiaceae/Ocimum/

3. *O. vanderystii*—Angola, Zaire, Zambia, Congo-Brazzaville
4. *O. vandenbrandei*—Zaire (Marungu Province)
5. *O. urundense*—Tanzania, Burundi
6. *O. tubiforme*—South African Northern Province
7. *O. transamazonicum*—Brazil
8. *O. tenuiflorum*—Holy Basil (Tulsi)—Indian subcontinent, China, Southeast Asia, Queensland, New Guinea; grow naturally in Fiji, Kenya, French Polynesia, Venezuela, West Indies
9. *O. spicatum*—Yemen, Ethiopia, Somalia, Kenya
10. *O. spectabile*—Ethiopia, Tanzania, Somalia, Kenya
11. *O. somaliense*—Ethiopia
12. *O. serratum*—South Africa, Switzerland
13. *O. serpyllifolium*—Somalia, Yemen, Saudi Arabia
14. *O. reclinatum*—KwaZulu-Natal, Mozambique
15. *O. pyramidatum*—Tanzania
16. *O. pseudoserratum*—South African Northern Province
17. *O. ovatum*—Paraguay, Brazil, Argentina, Uruguay
18. *O. obovatum*—Madagascar and tropical Africa
19. *O. nummularia*—Somalia
20. *O. nudicaule*—Paraguay, Brazil, Argentina (Misiones Province)
21. *O. natalense*—KwaZulu-Natal, Mozambique
22. *O. motjaneanum*—Switzerland
23. *O. monocotyloides*—Zaire
24. *O. mitwabense*—Zaire
25. *O. minutiflorum*—Central and Eastern Africa
26. *O. minimum*—Sri Lanka, India
27. *O. metallorum*—Zaire
28. *O. mearnsii*—Tanzania, Kenya, Uganda
29. *O. masaiense*—Kenya (Ngong Hills)
30. *O. lamiifolium*—Central and Eastern Africa
31. *O. labiatum*—South Africa, Mozambique, Swaziland
32. *O. kilimandscharicum*—Camphor basil—Tanzania, Kenya, Uganda, Ethiopia, Sudan; grow wild in India, Angola, Thailand, Myanmar
33. *O. kenyense*—Tanzania, Kenya
34. *O. jamesii*—Somalia, Ethiopia
35. *O. irvinei*—Western Africa
36. *O. hirsutissimum*—Zaire
37. *O. gratissimum*—African basil—Madagascar, Africa, Bismarck Archipelago, South Asia; naturally occurring in Mexico, Polynesia, Panama, Brazil, West Indies, Bolivia
38. *O. grandiflorum*—Kenya, Ethiopia, Tanzania
39. *O. fruticosum*—Somalia
40. *O. forskoelei*—East Africa from Egypt to Angola, Kenya, Arabian Peninsula
41. *O. formosum*—Ethiopian Bale Province
42. *O. fischeri*—Tanzania, Kenya
43. *O. fimbriatum*—Central Africa
44. *O. filamentosum*—Southern and Eastern Africa, Arabian Peninsula, Sri Lanka, India, Myanmar
45. *O. ericoides*—Zaire
46. *O. empetroides*—Zaire
47. *O. ellenbeckii*—Zaire, Ethiopia
48. *O. dolomiticola*—South African Northern Province
49. *O. dhofarense*—Oman
50. *O. decumbens*—commencing from Zaire to South Africa
51. *O. dambicola*—Zambia, Tanzania
52. *O. cufodontii*—Somalia, Ethiopia, Kenya

53. *O. coddii*—South African Northern Province
54. *O. circinatum*—Somalia, Ethiopia
55. *O. centraliafricanum*—Zambia, Zaïre, Zimbabwe, and Tanzania
56. *O. carnosum*—South America and Mexico
57. *O. canescens*—Tanzania
58. *O. campechianum*—Amazonian basil—prevalent across Mexico, Florida, West Indies, South and Central America
59. *O. burchellianum*—Cape Province in South Africa
60. *O. basilicum*—Sweet basil, basil—Indian subcontinent, China, Southeast Asia; found growing naturally in Africa, Russia, Mexico, Ukraine, South America, Central America, and different marine islands
61. *O. angustifolium*—Southeastern Africa from Kenya to Transvaal
62. *O. amicorum*—Tanzania
63. *O. americanum*—African tropics), Southeast Asia, China; grows naturally in Christmas Island, Queensland, and tropical parts of America
64. *O. africanum*—Indian subcontinent, China, Indochina, Madagascar, Africa; grown wild in Chiapas, Netherlands Antilles, Guatemala, Eastern Brazil

8.3 Some of the Important *Ocimum* Species and Their Medicinal Uses

8.3.1 *O. sanctum* L.

O. sanctum or *O. tenuiflorum* is a great revered curative plant in India and thus popular as 'holy basil.' It is identified as *Vishnupriya* (or Thulasi) in Sanskrit, *Kala-Thulasi* in Hindi, and Thulasi in Tamil. Holy basil is indigenous to India and regions of Eastern and Northern Africa, Taiwan, and Hainan Island and wildly grows all through India, up to Himalayas at an altitude of 1800 m (5900 feet) (Anonymous 1991; WHO 2002; Gupta et al. 2008). In China, it occurs in dry, sandy areas of Hainan and Sichuan, as well as in Indonesia, Cambodia, Laos, Myanmar, Malaysia, the Philippines, Vietnam, and Thailand (Li and Hedge 1994). Holy basil is cultivated in Southeast Asia and also grows abundantly in Australia, West Africa, and some Arab countries (Parotta 2001; Mondal et al. 2009). In India, the dried seed, dried leaf, and dried whole plant are separately used in the traditional medicine systems of *Unani*, *Siddha*, and *Ayurveda* as well as in Indian folk medicine (Ved and Goraya 2008). Mainly two types of *O. sanctum* L. are cultivated: (i) Green-leaved tulsi plants called *Sri Rama Tulsi* and (ii) Purple-leaved tulsi plants referred to as *Krishna/Shyama Tulsi* (Anonymous 1991).

8.3.1.1 Medicinal Potentialities and Other Uses

O. sanctum is extensively used by the conventional health practitioners to cure a variety of diseases such as bronchial asthma, bronchitis, malaria, dysentery, diarrhea, arthritis, skin diseases, painful eye diseases, insect bite, chronic fever. It is attributed to have anticancer, antifertility, antidiabetic, antimicrobial, antifungal, antiemetic, cardioprotective, antispasmodic, hepatoprotective, analgesic, adaptogenic, and diaphoretic actions (Prakash and Gupta 2005). It has also been reported that upon consuming the fresh leaves of holy basil with milk or tea helps in curb acidity, heart burn, and vomiting. The paste of powder of the dried leaves and mustard oil helps to cure tooth troubles like pyorrhea and foul smell. The seeds of the plant are used in the treatment of genitourinary system disorders (Bhasin 2012).

8.3.1.2 Essential Oil

Holy basil possesses a strong taste, somewhat pungent as compared to other basils due to the presence of β-caryophyllene (a sesquiterpenoid), and eugenol (a phenylpropanoid) in the leaf. The leaf essential oil is used to relieve the joint pains, body rashes, and inflammation (Bhasin 2012). Research studies on the essential oils of different *O.* species have revealed that its oil owes biological activities. Of these, antibacterial, antifungal, and antimicrobial properties are crucial (Kalita and Khan 2013).

High-quality essential oil isolate, eugenol, is the main component of the oil, which is of immense importance in food and flavoring industry. Eugenol serves as a source of vanillin synthesis, which is the world's most popular flavor in all sorts of food products. Eugenol is of great value in pharmaceutical trade and is by and large obtained from bud of cloves (70–85%) and from *Cinnamomum* leaves/barks (20–50%) (Mukherji 1987). Though these plant sources are eugenol-rich, but industrial extraction of eugenol from them is a costly affair. On the contrary, *O. sanctum L.* (Tulsi) is an economical source to extract eugenol commercially (Mukherji 1987). The volatile oil from leaf also contains pienene (α- and β), linalool, carvacrol, camphene, limonene, sabinene, β-elemene, β-caryophyllene, cineole, methyl chavicol, germacrene D, and ursolic acid. (Lal et al. 2003; Pattanayak et al. 2010). Variations in the quantities of eugenol were observed in the essential oil profile of *O. sanctum* L. growing in diverse parts of India (Mukherji 1987; Rajeshwari 1992). Eugenol percent in the essential oil composition of *O. sanctum* L. ranges between 40 and 70%. During the monsoon season in India, usually a significant decrease of eugenol content in the leaf of many *O. sanctum* genotypes was observed, and at times, the decrease is up to 0% from 80%.

8.3.2 *O. gratissimum* L.

O. gratissimum is an aromatic, perennial herb that naturally originated in tropical areas of India, but is also reported from West Africa. It occurs in Savannah and coastal areas of Nigeria. Along with Ceylon, the South Sea Islands, it is also cultivated in Nepal, Deccan, Bengal, and Chittagong (Nadkarni 1999). In its native area, *O. gratissimum* emerge up to 150 m altitude from sea level in the coastal areas, along the shore of lakes, in the submontane forest, savanna vegetation and deranged land. It is cultivated as the hedge plant in Southeast Asia, up to an altitude of about 30 m. It is known by different names like wild basil, tree basil or clove basil in various parts of the world, but in India, ithas many vernacular names, the most frequently used names are *Ram tulsi* (Hindi), *Vriddhutulsi* (Sanskrit), and*Nimma tulasi* (Kannada).

8.3.2.1 Medicinal Potentialities and Other Uses

O. gratissimum has considerably been used in the conventional medical systems all around the globe. The plant possesses a sharp and pungent taste, but is effective in curing diseases related to brain, liver, heart, and spleen; in addition, it also strengthens the gums and makes a person get rid of foul breath. It is stomachic, diaphoretic, laxative and is one of the good treatments of fever (Bhasin 2012). The whole plant preparations are used to treat stomach and headaches, sunstroke, and influenza (Orwa et al. 2009). It is exploited for medicinal, condiment, and culinary purposes in Northeastern Brazil. Plentiful with essential oils, the flowers and the leaves are also used to prepare teas and infusion (Rabelo et al. 2003). In the areas of Savannah, leaf decoctions are used against mental sickness (Akinmoladun et al. 2007). Ibos people in the southeastern Nigeria use *O. gratissimum* in managing the baby's cord, as well as for keeping the sterility of the wound surfaces. Inhabitants of Brazilian tropical forest prepare a decoction of *O. gratissimum* roots as a tranquilizer to calm down children (Cristiana et al. 2006). In addition, it is also helpful in curing fungal infections, cold, fever, and catarrh (Ijeh et al. 2005). Apart from antimicrobial and antifungal antioxidant activities, *O. gratissimum* also has pharmacological properties like antidiarrhoeal, anti-inflammatory, cardiovascular, immunostimulatory, antidiabetic, hepatoprotective, wound healing, antihypertensive effects as well as analgesic and anticonvulsant activities (Prabhu et al. 2009).

8.3.2.2 Essential Oil

About 0.8–1.2% essential oil is present in the fresh aerial parts of *O. gratissimum,* which is pale yellow in color and accommodates high eugenol percentage. Orwa et al. (2009) have reported the variable chemical composition of the oils from six chemotypes which were characterized by some of the major components like eugenol, citral, thymol, linalool, geraniol, and ethyl cinnamate in the essential oil. But there lacks a general idea related to the occurrence of its diverse types and possible repercussion for taxonomy. Jirovetz et al. (2003) while performing the chemotaxonomic analysis of the aroma compounds present in the essential oil of *Ocimum* species occurring in southern India found that the leaf essential oil of *O. gratissimum* includes eugenol (63.36%) in the highest amount followed by (Z)-β-ocimene (9.11%), germacrene D (8.84%), and β-caryophyllene (3.89%). The essential oil is used in food product flavoring, preparation of beverages and detergents, dental health care as well as mosquito repellant moreover to its antibacterial and antifungal properties (Bhasin 2012). The essential oil can also be put on to combat fever, throat, ear or eye inflammations, diarrhea, stomach pain, and skin disease (Orwa et al. 2009).

8.3.3 *O. viride*

O. viride, popularly known to be 'Van Tulsi,' is an aromatic soft-growing herb which grows annually. It is the native of West Africa, also introduced into India. It is used throughout Africa for fevers including malaria (http://www.richters.com/Web_store/web_store.cgi?product=X1292&show=&prodclass=Herb_and_Vegetable_Plants&source=799158.29456). It is called as 'Fever-plant' at Sierra Leone, and its decoction is drunk as tea. Due to the febrifugal property of the leaves, it is famous as Taap-maari Tulasi in Indian folks of Maharashtra (http://www.springerreference.com/docs/html/chapterdbid/69023.html). Being perennial in nature, the stem of the plant is more or less erect, profusely branched, with brownish green elliptic lanceolate leaves. Flowers are light yellow. Seeds are brownish, globose, and non-mucilaginous (Bhasin 2012).

8.3.3.1 Medicinal Potentialities and Other Uses

The plant is considerably used as a fomentation for lumbago and rheumatism. Leaf decoction is used against fever and cough. Fresh leaf juice of the plant is used in catarrh and as eye drops for conjunctivitis (Anonymous 1966). The use of spicy herbs like basil in food not for the only purpose of flavoring, rather they serve some medicinal purposes also, such as anti-inflammatory, antioxidant, antimicrobial, or antiviral. In the countries like Ghana, fresh basil is used in food preparations (such as chicken sauces, soups) as well as some local purposes. Since the plant is scant during the off-seasons, it requires skilled preservation. Moreover, the plant is highly delicate and has to be protected and conserved against decaying and spoilage (Danso-Boateng 2013).

8.3.3.2 Essential Oil

Essential oil is viscous, yellowish in color, having a distinctive smell of thymol with a spicy and pungent flavor. It is widely used in fragrance, flavor, and medicated products and as a potent antioxidant, antiseptic, disinfectant, and preservative. It is also utilized in synthetic essential oil compounding, in addition as a preparatory material for manufacturing menthol synthetically (Bhasin 2012). Being an antiseptic, high thymol content in leaf oil describes its use in mouthwash, wound dressing, and treatment for conjunctivitis (http://www.richters.com/Web_store/web_store.cgi?product=X1292&show=&prodclass=Herb_and_Vegetable_Plants&source=799158.29456). The other components of essential oil were d-Limonene, terpineol, α- and ϒ-terpinene

8.3.4 *O. basilicum* L.

The sweet basil (*O. basilicum*), with its great flavor and aroma, stands to be an extremely popular and extensively cultivated herbs all around the world. Several name origins and viewpoints are connected with basil, but the widespread name of basil is most likely known to be derived from the Greek words, *basileus* which means 'king' or *basilikon* signifying 'royal.' It is a low and soft-growing herb, initially native to India, China, Southeast Asia, and New Guinea (http://en.wikipedia.org/wiki/Basil). It was originally cultivated in India for over 5000 years. In India, it is vernacularly popular as Indian basil, Ram Tulsi (Hindi), Sabje (Gujarati), Sabza (Marathi), Tirunittru (Tamil), Khubkalam (Bengali), and Kattu tulasi (Malayalam) (http://www.flowersofindia.net/catalog/slides/Basil.html). It is 20–60 cm in height, with light green, opposite, silky leaves, 1.5–5.0 cm in length, and 1–3 cm in breadth. More or less it feels like anise, with a sharp, spicy, and sweet aroma. Sweet basil is very much cold susceptible, with the finest growth in the dry and hot environment (http://en.wikibooks.org/wiki/Horticulture/O._basilicum). Basil is a crop of great economic significance which is capable of producing 100 tonnes of essential oil annually throughout the world and with a commerce value of around US$15 million as a potherb per year. It is also extensively used in traditional medicinal systems (http://www.kew.org/science-conservation/plants-fungi/O.-basilicum-basil).

8.3.4.1 Medicinal Potentialities and Other Uses

Sweet basil is a commonly cultivated fragrant crop which is grown either for essential oil production, dry leaves in the fresh market, or for ornate purposes. Conventionally, sweet basil has also been used as a curative plant for treating coughs, headaches, constipation, diarrhea, warts, worms, and kidney malfunctions. It is also considered to be an antispasmodic, stimulant, stomachic, carminative, and insect repellent (Simon et al. 1990). In general, the plant also treats problems related to nervous system and digestion. Tea which is made from the fresh leaves is believed to prevent headache, nausea, and mild nervous tensions. Basil leaves boiled in water help in curing sore throat. Leaf decoction stands as a good medication for treating respiratory ailments. Basil leaf juice promotes kidney stones ejection. Stress, mouth infections, and ulcer could be prevented if basil leaves are chewed daily. The plant is also helpful in reducing the blood cholesterol (http://health-from-nature.net/Basil.html).

8.3.4.2 Essential Oil

Oil is light yellow in color, slightly thick in texture having a sweet herbal spicy aroma, with methyl chavicol as a key ingredient. The oil is used in aroma and flavor industry (Bhasin 2012). It is widely used in the food, cosmetic, aromatherapy, and medicinal industries because of its pleasant fragrance, in addition to the antimicrobial and insecticidal activities (Zheljazkov et al. 2008). Typically, the basil essential oil contains high amounts of estragole (methyl chavicol) and linalool. Other components of the essential oil that are present in low concentrations are: eugenol, 1,8-cineole, beta-caryophyllene, alpha-terpineol, sabinene, geraniol, gamma-terpinene, alpha-phellandrene, thujone, limonene, ocimene, myrcene, and para-cymene. Volatile ingredients in sweet basil are reported to affect the distribution, composition, and spore germination of a few fungal populations (Simon et al. 1990).

8.3.5 *O. americanum* L.

Limehairy or 'hoary basil' (*O. americanum*) is an annual herb which possesses lavender- or white-colored flowers. It is employed for therapeutic purposes and is commonly known as *Kali Tulsi* in India. Regardless of the ambiguous name, it is indigenous to Africa, China, the Indian subcontinent, and Southeast Asia. The species grows wild in Christmas Island, Queensland, and parts of American tropics (http://en.wikipedia.org/wiki/O._americanum). Lime basil possesses ovoid leaves having the tender scalloping margins with deep veins which are bright and light green in color. They are arranged opposite on square stems. The flowers bloom from July onwards at the branch tops in the large spikes with small white flowers. Plants usually reach the height of up to three feet and the width is about 18″. Under optimum conditions, the flower spikes attain good height (http://www.floralencounters.com/Seeds/seed_detail.jsp?productid=92647).

8.3.5.1 Medicinal Potentialities and Other Uses

The plant is carminative, stimulant, and diaphoretic and used against cold, cough, catarrh, and bronchitis. Fresh leaf juice is beneficial against toothache when used as mouthwash and also fetches relief in dysentery; putting the leaf juice drops into nostrils helps curing migraine. Parasitical skin diseases can also be cured using the basil leaf paste. Tea prepared from the leaves is used as a remedy for fever, indigestion, and diarrhea. Fumes of the dried plant when burnt serve as a good mosquito repellent. The essential oil from the leaves and inflorescences is attributed to antimicrobial properties (http://www.mpbd.info/plants/ocimum-americanum.php). Sunitha and Begum (2013) recently reported that the methanolic extract of *O. americanum* L. seeds possesses promising immunomodulatory activity (Bhasin 2012).

8.3.5.2 Essential Oil

Aerial parts yield the pale yellow-colored essential oil, marked by a distinct citrus smell due to the presence of citral being its main constituent accompanied by other components like citronellal, methyl cinnamate, citronellic acid, eugenol, methyl heptenone, and geraniol in addition to terpinolene, dipentene, crithmene, pinene, caryophyllene, limonene, camphene, sabinene as well as phenol and acetic acid in traces (http://www.mpbd.info/plants/ocimum-americanum.php). The oil expansively finds a huge application in aroma, flavor, and therapeutic trade. In addition to producing vital fragrance grade *B*-ionone isolates, it is also used in the vitamin A synthesis (Bhasin 2012).

8.3.6 *O. canum* Sims.

Native of the African continent, Basilic camphor (*O. canum Sims*) is a local south Indian species. The plant being an annual herb attains a height of 2 feet with ovate leaves and light pinkish flowers. Mucilaginous and dark brown in color, the seeds are narrowly ellipsoid. Also recognized as the African basil, it possesses a distinctive flavor of mint and acquires scented flowers including hairy leaves. In particular, *O. canum* is used as a remedy against diabetes. Introduced lately in the Americas, it was common in areas of the tropics as well as tropical Africa. Chemotypic varieties, containing linalool, eugenol, estragole, camphor, methyl chavicol, methyl cinnamate, β-caryophyllene, trans-α-bergamotene, α-terpineol, germacrene D, thymol, fenchone, *p*-cymene, and 1,8-cineol as major components, have been reported in India, East Asia, Europe, North, Central and West Africa, and America (Chagonda et al. 2000).

8.3.6.1 Medicinal Potentialities and Other Uses

Besides the culinary value, *O. canum* is highly beneficial in curing various types of ailments, especially lowering blood glucose levels in type 2 diabetes. In addition, the herb is used to cure fevers, colds, dysentery, tooth problems, headaches, parasitic infestations on the body, and joint inflammation. The traditional medicine recognized its value in the treatment of fevers, dysentery, and tooth problems. Being an insect repellent, it is also used to combat the post-harvest insect damages. Linalool is the main essential oil constituent of *O. canum*. The seeds are a source of dietary fiber which helps in reducing constipation (http://www.agriculturalproductsindia.com/aromatic-plants/aromatic-plants-ocimum-canum.html).

8.3.6.2 Essential Oil

Pale yellow in color, the essential oil is a rich source of linalool and has been reported to have strong antibacterial and antifungal activities (Bhasin 2012). Therefore, it is of more commercial use for fragrance, flavor, and cosmetic industries like in chewing gums, teas, sweets, milk products, soft drinks, energy drinks, soaps, shampoos, body lotions, shower gels, and toothpastes. It works as a remarkable mosquito repellent and thereby contributes in preventing chronic fevers like malaria and dengue (http://www.agriculturalproductsindia.com/aromatic-plants/aromatic-plants-ocimum-canum.html).

8.3.7 *O. carnosum* LK. et. Otto.

The perennial *O. carnosum* or *O. selloi* ('basil pepper') is indigenous to South America, and there it is popular as 'elixir paregorigo,' 'anis,' 'alfavacadeanis,' and 'alfavaquinha.' The plant is an under the shrub which is much branched having dark green-colored and simple ovate-oblanceolate leaves. Flowers are purplish in color but are small. Seeds are slightly mucilaginous, ellipsoid in shape, and purplish to dark brown in color.

8.3.7.1 Medicinal Potentialities and Other Uses

The plant has antipyretic, carminative, antidiarrheal, antispasmodic, analgesic, anti-inflammatory, and antibacterial properties.

8.3.7.2 Essential Oil

Pale yellow in color, the essential oil is viscid accompanied by an aroma of earthy zest due to the presence of one of the oil components, elemicin. Expensive by cost, elemicin possesses immense medicinal and flavoring characteristics. The manufacture of 3,4,5-trimethoxy-benzaldehyde which is the substrate for the synthesizing trimethoprim (a constituent employed to produce the antibacterial drug, Septran) requires elemicin for its production (Bhasin 2012). Conventionally, the essential oil has been used as insect repellant, in perfumery, food flavoring, in traditional medicine and as spice. *O. selloi* in Brazil is used to soothe inflammations and stomachaches. In 2003, an investigation carried out at the State University of Ponta Grossa, Brazil, revealed that oil of green pepper basil is an efficient mosquito repellent and also does not leave any irritation to the skin of humans (http://www.motherearthliving.com/plant-profile/green-pepper-basil.aspx#axzz3CWsYnJot). Anethole and methyl chavicol serve as the key constituents of the essential oil obtained from flowers, leaves, and stem. *O. selloi* essential oil has unpretentious antimicrobial activity against bacterial strains of *Staphylococcus aureus* and *Escherichia coli, and* it has an efficient repellent activity for the *Anopheles braziliensis* mosquito (Gonçalves et al. 2010).

8.3.8 *O. kilimandscharicum* Guerk.

O. kilimandscharicum Guerke ('Camphor Basil'), called as 'Kapoori Tulsi' in Hindi, is the plant of foreign origin (Africa), but is also grown in hilly as well as plain areas of south India and some parts of Turkey. Being a source of camphor, this plant became a center of attraction (Gill et al. 2012). This species of basil acquires a sharp yet less pleasing flavor. The plant is described to be perennial, possessing leaves that are simple, ovate, and oblong. Seeds are mucilaginous, ovoid to oblong, and black to brown in color.

8.3.8.1 Medicinal Potentialities and Other Uses

The plant has many therapeutic properties like carminative, antipyretic, stimulant, antibacterial, and antifungal. In East African tradition, the extracts of *O. kilimandscharicum* were used for bringing ease to many health disorders such as—treatment of diarrhea, coughs, colds, measles, and abdominal pains. It is also reported to be a good insect repellent, specifically hostile to mosquitoes and as a storage pest control (Kokwaro 1976). Camphor has a good toxicity and protectant potential and is the major constituent of the plant; hence, it is reported to be critical of product beetles (Ofori et al. 1998). Scientific evaluations carried out on this plant's curative and insecticidal efficiency categorize it as an aromatic plant which finds good applications in medical, aroma therapy, and pesticide commerce. The plant leaves are aromatic and documented to have insecticidal, antibacterial, antiviral, thermogenic, acrid, ophthalmic, appetizing, and deodorant properties (Gill et al. 2012).

8.3.8.2 Essential Oil

Essential oil is marked with the strong smell of camphor and is pale yellow in color. The oil is applicable against diarrhea, sprains, and dental/oral preparations. Also, it is enormously used in pharmaceutical, aroma, and flavor industries (Anonymous 1966; Bhasin 2012). The essential oil profile also reveals camphor (60–80%) to be its chief constituent, while other components are camphene, 4-terpineol, limonene, 1,8-Cineole, α-terpineol, linalool, trans-caryophyllene, myrtenol, endo-borneol, etc. (http://www.infonet-biovision.org/default/ct/193/medicinalPlants).

8.3.9 *O. suave* Willd.

O. suave is a tropical shrub, the native of Africa and India. Its habitat is from Guinea to West Cameroons and generally across Africa to East Africa into tropical Asia and is common in the upland forest areas of East Africa (http://www.fao.org/docrep/x2230e/x2230e09.htm). It is an annual plant which needs protection from frost. Straight and upright, this African bush basil gets woody with time. The seed heads of *O. suave* are reddish in color and seem to be arching throughout the wild lands within the local areas of its cultivation (https://www.horizonherbs.com/product.asp?specific=2280).

8.3.9.1 Medicinal Potentialities and Other Uses

The plant is used as a traditional medicine for treating stomachache, nasal congestion, cough, influenza, and eye/ear inflammation. It is also used as a perfume, an insect repellent (particularly against mosquitoes), and a grain protectant (http://www.fao.org/docrep/x2230e/x2230e09.htm).

8.3.9.2 Essential Oil

Pale yellow-colored essential oil is greatly viscid, having dusty wood balsamic odor. Since the oil has abundant sesquiterpene alcohols, it serves to be a good source of tobacco and slay flavoring, as well as an ingredient in mosquito repellent and body perfumes (Bhasin 2012). The oil showed repellent properties against *S. zeamais* when assessed 1 h after application in an olfactometer (http://www.fao.org/docrep/x2230e/x2230e09.htm). Other constituents of oil include eugenol and mono- and sesquiterpenoids.

8.4 Advantages of *Ocimum* Cultivation

- Employment generation due to developing subsidiary trade.
- Limits the rural populace relocation to metropolitan areas through creation of rustic employment.
- Great gross profits in comparison with conventional horticultural and agricultural crops.
- Could easily be incorporated in the existing system of farming/cropping.

Tulsi cultivation finds both religious and functional importance connecting the cultivator with inventive vigor of environment, in addition, offering way-out against hunger, food insecurity, rural poverty, climate change, and environmental depletion. Allopathic drug market worldwide has accomplished the mark of US$500 billion, though the international trade involving the plant-based medicines was anticipated to be US $100 billion which is likely to rise up to US$5 trillion by 2050; hence, tulsi (basil) plants are progressively being admired as a significant source of livelihood opportunities for rustic poor as well as a revenue source for the government also. Growing the collection of Tulsi (basil) thus presents to be an eminent source of financial earning to the rural population, particularly women, forest-reliant primeval tribes, deprived and marginalized farmers. As per Indian government statistics, the processing and collection of Medicinal and Aromatic Plants (MAPs) put in a minimum of 35 million working days of employment annually (Planning Commission 2000). The predicted global demand of MAPs is 60–62 billion US $ which is escalating per year at a rate of 7–10%. Due to poor irrigation, strewn farms, animal attack, financial difficulties, and high input cost, the producers are hesitant to bring their produce by themselves in the market. Even though the companies work at the village level, the government must contribute for basil cultivation and reinforcing its marketing. Provisions should be made by the government regulatory authorities related to market for making available the loan to farmers at a low rate of interest. This would certainly prove appealing to poor farmers and would also raise confidence and support cultivation of the basil crops. In an effort to provide focus for raising the livelihood for thousands of farmers, the farming of tulsi goes ahead of providing profit to folks and family and has commenced to attend to wider economic, social, and environmental issues.

References

Akinmoladun AC, Ibukun EO, Emmanuel A, Obuotor EM, Farombi EO (2007) Phytochemical constituent and antioxidant activity of extract from the leaves of *Ocimum gratissimum*. Sci Res Essay 2:163–166

Anonymous (1966) The wealth of India—raw materials, vol III. CSIR, New Delhi

Anonymous (1991) The wealth of India. Publication and Information Directorate, CSIR, New Delhi, pp 79–89

Bhasin M (2012) *Ocimum*—taxonomy, medicinal potentialities and economic value of essential oil. J Biosph 1:48–50

Chagonda LS, Makanda CD, Chalchat J (2000) The essential oils of *Ocimum canum* Sims (basilic camphor) and *Ocimum urticifolia* Roth from Zimbabwe. Flav Fragr J 15:23–26

Commission Planning (2000) Report of the taskforce on medicinal plants in India. Planning Commission, Government of India, Yojana Bhawan, New Delhi

Cristiana M, Murbach F, Márcia OM, Mirtes C (2006) Effects of seasonal variation on the central nervous system activity of *Ocimum gratissimum* L. essential oil. J Ethnopharmacol 105:161–166

Danso-Boateng E (2013) Effect of drying methods on nutrient quality of Basil (*Ocimum viride*) leaves cultivated in Ghana. Int Food Res J 20(4):1569–1573

Darrah HH (1980) The cultivated basils. Buckeye Printing Company, Independence

Deschamps C, Simon JE (2010) Phenylpropanoid biosynthesis in leaves and glandular trichomes of basil (*Ocimum basilicum* L.). Meth Mol Biol 643:263–273

Gang DR, Simon J, Lewinsohn E, Pichersky E (2002) Peltate glandular trichomes of *Ocimum basilicum* L. (Sweet Basil) contain high levels of enzymes involved in the biosynthesis of phenylpropenes. J Herbs Spices Med Plants 9(2):189–195

Gill D, Soni N, Sagar B, Raheja S, Agrawal S (2012) *Ocimum kilimandscharicum*: a systematic review. J Drug Deliv Ther 2(3):45–52

Gonçalves L, Azevedo AA, Otoni WC (2010) Characterization and ontogeny of the glandular trichomes of *Ocimum selloi* Benth. (Lamiaceae). Acta Bot Bras 24 (4):909–915 (Feira de Santana)

Gupta SM (1971) Plant myths and traditions in India. E. J. Brill, Leiden, Vangsgaards Antikvariat Aps (Copenhagen, Denmark), pp 66–72

Gupta AK, Tandon N, Sharma M (eds) (2008) *Ocimum sanctum* Linn. In: Quality standards of Indian medicinal plants, vol 5. Medicinal Plants Unit, Indian Council of Medical Research, New Delhi, pp. 275–284

Ijeh II, Omodamiro OD, Nwanna IJ (2005) Antimicrobial effects of aqueous and ethanolic fractions of two spices, *Ocimum gratissimum* and *Xylopia aethiopica*. Afr J Biotechnol 4:953–956

Jirovetz L, Buchbauer G, Shafi MP, Kaniampady MM (2003) Chemotaxonomical analysis of the essential oil aroma compounds of four different *Ocimum* species from southern India. Eur Food Res Technol 217:120–124

Kalita J, Khan ML (2013) Commercial potentialities of essential oil of *Ocimum* members growing in North East India. Int J Pharm Life Sci 4(4):2559–2567

Kokwaro JO (1976) Medicinal plants of East Africa. East African Literature Bureau, Kul Graphics Ltd., Nairobi

Lal RK, Khanuja SPS, Agnihotri AK, Misra HO, Shasany AK, Naqvi AA, Dhawan OP, Kalra A, Bahl JR, Darokar MP (2003) High yielding eugenol rich oil producing variety of *Ocimum sanctum*—CIM-Ayu. J Med Arom Plant Sci 25:746–747

Li XY, Hedge IC Lamiaceae (1994). In: Wu ZY, Raven PH (eds) Flora of China, vol 17 (Verbenaceae through Solanaceae). Science Press, Beijing and Missouri Botanical Garden Press, St. Louis, pp 296–297

Meyers M (2003) Basil: an herb society of America guide. The Herb Society of America, Kirtland

Mondal S, Mirdha BR, Mahapatra SC (2009) The science behind sacredness of tulsi (*Ocimum sanctum* Linn.). Indian J Physiol Pharmacol 58(4):291–306

Mukherji SP (1987) *Ocimum*—a cheap source of Eugenol. Sci Rep 599

Nadkarni KM (1999) Indian materia medica, 3rd edn. Popular Prakashan Pvt Ltd, G. G. Pathare at Popul"r Press (Born.) Ltd., Bombay and Published by G. R. Bhatkal, jointly for the Popular Book Depot (Regd.), Bombay and Dhootapapeshwar Prakashan Ltd., Panvel

Ofori DO, Reichmuth CH, Bekele AJ, Hassanali A (1998) Toxicity and protectant potential of camphor, a major component of essential oil of *Ocimum kilimandscharicum*, against four-stored product beetles. Int Pest Manag 44(4):203–209

Orwa C, Muta A, Kindt R, Jamnads R, Simons A (2009) Agrofrestre database: a tree reference and selection guide version 4.0. http://www.orldagrofrestry.org/aftredb/

Parotta JA (2001) Healing plants of peninsular India. CABI Publishing, New York

Pattanayak P, Behera P, Das D, Panda SK (2010) *Ocimum sanctum* Linn. A reservoir plant for therapeutic applications: an overview. Pharmacog Rev 4(7):95–105

Prabhu KS, Lobo R, Shirwaikar AA, Shirwaikar A (2009) *Ocimum gratissimum*: a review of its chemical, pharmacological and ethnomedicinal properties. Open Complement Med J 1:1–15

Prakash P, Gupta N (2005) Therapeutic uses of *Ocimum sanctum linn* (tulsi) with a note on eugenol and its pharmacological actions: a short review. Indian J Physiol Pharmacol 49(2):125–131

Rabelo M, Souza EP, Soares PMG, Miranda AV, Matos FJA, Criddle DN (2003) Antinociceptive properties of the essential oil of *Ocimum gratissimum* L. (Labiatae) in mice. Braz J Med Biol Res 36:521–524

Rajeshwari S (1992) *Ocimum sanctum*. The Indian home remedy. In: Rajeshwari S (ed) Current medical scene. Rajeswari Foundations Limited, Bombay

Simon JE, Quinn J, Murray RG (1990) Basil: a source of essential oils. In: Janick J, Simon JE (eds) Advances in new crops. Timber, Portland, pp 484–989

Sunitha K, Begum N (2013) Immunomodulatory activity of methanolic extract of *Ocimum americanum* seeds. Int J Res Pharm Chem 3(1):95–98

Thiselton-Dyer TF (1889) The folk-lore of plants. D. Appleton and Company, New York

Ved DK, Goraya GS (2008) Demand and supply of medicinal plants in India. Bishen Singh Mahendra Pal Singh, Dehradun

World Health Organization (2002) Folium Ocimi Sancti. WHO monographs on selected medicinal plants, vol 2. World Health Organization, Geneva, pp 206–216

Zheljazkov VD, Callahan A, Cantrell CL (2008) Yield and oil composition of 38 Basil (*Ocimum basilicum* L.) accessions grown in Mississippi. J Agric Food Chem 56:241–245

9 Triterpene Functional Genomics in *Ocimum*

Sumit Ghosh

Abstract

Triterpenes, isoprene-derived 30-carbon compounds, constitute a large class of natural products with enormous structural and functional diversity. Bioactive triterpenes are used as ingredients in pharmaceuticals, foods, and cosmetics. *Ocimum* species are known to produce bioactive triterpenes such as ursolic acid, oleanolic acid, betulinic acid, alphitolic acid, euscaphic acid, and epi-maslinic acid. Functional genomics studies conducted in *O. basilicum* led to identification and characterization of oxidosqualene cyclases (OSCs) and cytochrome P450s (P450s) for the biosynthesis of ursolic acid and oleanolic acid, and opened up the prospects for producing bioactive triterpenes in alternate microbial hosts such as yeast. Moreover, a recent advancement in the area of high-throughput sequencing of genomes and transcriptomes provided an opportunity to understand the molecular and biochemical basis for the biosynthesis of diverse triterpenes and other phytochemicals in *Ocimum* species.

S. Ghosh (✉)
Biotechnology Division, CSIR-Central Institute of Medicinal and Aromatic Plants, Lucknow 226015, India
e-mail: sumitghosh@cimap.res.in; sumitg80@gmail.com

9.1 Introduction

Triterpenes are the isoprene-derived 30-carbon compounds. Triterpenes isolated from the natural sources displayed enormous structural and functional diversity (Hill and Connolly 2017). The commercial applications of triterpenes in the pharmaceutical, food, and cosmetic sectors were attributed to their potent bioactivities such as anticarcinogenic, hepatoprotective, anti-inflammatory, antioxidant, and antimicrobial (Sawai and Saito 2011; Sheng and Sun 2011; Moses et al. 2013). Understanding triterpene biosynthesis is of particular interest for improving the yield of rare triterpenes in their natural hosts and for developing alternate sources for the sustainable production of triterpenes (Arendt et al. 2017; Reed et al. 2017).

Ocimum species such as *O. basilicum*, *O. tenuiflorum*, *O. gratissimum*, *O. americanum*, *O. micranthum*, *O. kilimandscharicum,* and *O. selloi* were found to produce bioactive triterpenes. In *O. basilicum*, in-depth studies were conducted to understand the biosynthesis and spatiotemporal accumulation of bioactive triterpenes such as ursolic acid and oleanolic acid (Misra et al. 2014, 2017). *O. basilicum* genes/enzymes for the biosynthesis of ursolic acid and oleanolic acid were identified and functionally characterized. A set of oxidosqualene cyclases (OSCs) and cytochrome P450s (P450s) of *O. basilicum* was found to be involved in the formation of ursolic

A. K. Shasany and C. Kole (eds.), *The Ocimum Genome*, Compendium of Plant Genomes, https://doi.org/10.1007/978-3-319-97430-9_9

acid and oleanolic acid from 2,3-oxidosqualene, the general triterpene precursor. This chapter started with an overview of triterpene structural diversity and classification and further discussed functional genomics studies directed toward understanding the biosynthesis of bioactive triterpenes in *Ocimum*. This chapter also highlighted recent advances in the area of high-throughput sequencing of genomes and transcriptomes, and establishment of the functional genomics tools such as transgenics, RNA interference (RNAi), virus-induced gene silencing (VIGS), hairy root culture in *Ocimum* species that will be useful in understanding the biosynthesis of diverse triterpenes and other phytochemicals in *Ocimum* species.

9.2 Triterpene Structural Diversity and Classification

Triterpenes are extremely diverse structurally, ranging from acyclic to hexacyclic, although tetracyclic and pentacyclic scaffolds are more abundant in nature (Fig. 9.1). More than 20,000 triterpene structures originating from over 100 scaffold types are known from the natural sources. Triterpene scaffolds are modified with various functional groups such as hydroxyl, carboxyl, carbonyl, alkyl, malonyl, and glycosyl, leading to a wide spectrum of biological activities. On the basis of the structural backbones, triterpenes are classified as squalene derivative, cucurbitane, cycloartane, dammarane, euphane, friedelane, gammacerane, holostane, hopane, isomalabaricane, lanostane, lupane, oleanane, protostane, quassinoid, saponin, serratane, tetranortriterpene, tirucallane, ursane (Hill and Connolly 2017). The diverse types of triterpenes are usually originated from two acyclic triterpenes, squalene, and 2,3-oxidosqualene (squalene epoxide), in prokaryotes and eukaryotes, respectively. Moreover, 2,3-oxidosqualene is also utilized for the biosynthesis of steroidal scaffolds in plants, animals, insects, and in a few microbes (Thimmappa et al. 2014; Ghosh 2016).

Sapelenins, callicarpol, squalene-1,10,24,25,30-pentol, cupaniol, and lobophytene isolated from *Entandrophragma cylindricum, Callicarpa macrophylla, Rhus taitensis, Cupania latifolia,* and *Lobophytum* sp., respectively, represent the acyclic squalene derivative class (Hill and Connolly 2013). However, cyclic triterpenes with one to six ring structures are more commonly known in plants. For example, achilleol A and camelliol C are the monocyclic triterpenes of *Achillea odorata* and *Camellia sasanqua*, respectively (Xu et al. 2004). Bicyclic triterpenes include myrrhanones, (+)-myrrhanol C, lamesticumins, and lansic acid which were isolated from *Commiphora mukul, Pistacia lentiscus,* and *Lansium domesticum*. Tricyclic triterpenes kadcotriones, achilleol B and schiglautone A, were found in a few plants such as *Kadsura coccinea, A. odorata,* and *Schisandra glaucescens* (Meng et al. 2011; Liang et al. 2013). Lanostane, cucurbitane, holostane, protostane, dammarane, euphane, and tirucallane are different classes of tetracyclic triterpenes (Fig. 9.1). Tetracyclic triterpenes are widespread in plant species and constitute major bioactive components in cucurbit and panax species (Hill and Connolly 2013; Shang et al. 2014). Triterpenes of the friedelane, gammacerane, hopane, lupane, oleanane, serratane, taraxastane, and ursane classes have pentacyclic scaffolds. Among these pentacyclic classes, lupane, oleanane, and ursane are widely distributed in the plant kingdom and represent major pentacyclic triterpene classes. Bioactive pentacyclic triterpenes are found in *Ocimum* sp., *Glycyrrhiza* sp., *Bacopa monnieri, Olea europaea, Terminalia* sp., *Taraxacum officinale, Rosmarinus officinalis* (Ghosh 2016). Hexacyclic triterpenes, although rare in nature, were also isolated from plants such as *Euscaphis japonica, Paeonia suffruticosa,* and *Kadsura longipedunculata* (Pu et al. 2005; Li et al. 2016). Quassinoids having C_{18}, C_{19}, C_{20}, C_{22}, and C_{25} skeletons are degraded triterpenes. Quassinoids are common in plants of the *Simaroubaceae* family. However, tetranortriterpenes are highly oxygenated triterpenes, commonly known in citrus and neem (Roy and Saraf 2006). Tetranortriterpenes have a prototypical structure, either containing or derived from a precursor with a 4,4,8-trimethyl-17-furanylsteroid skeleton.

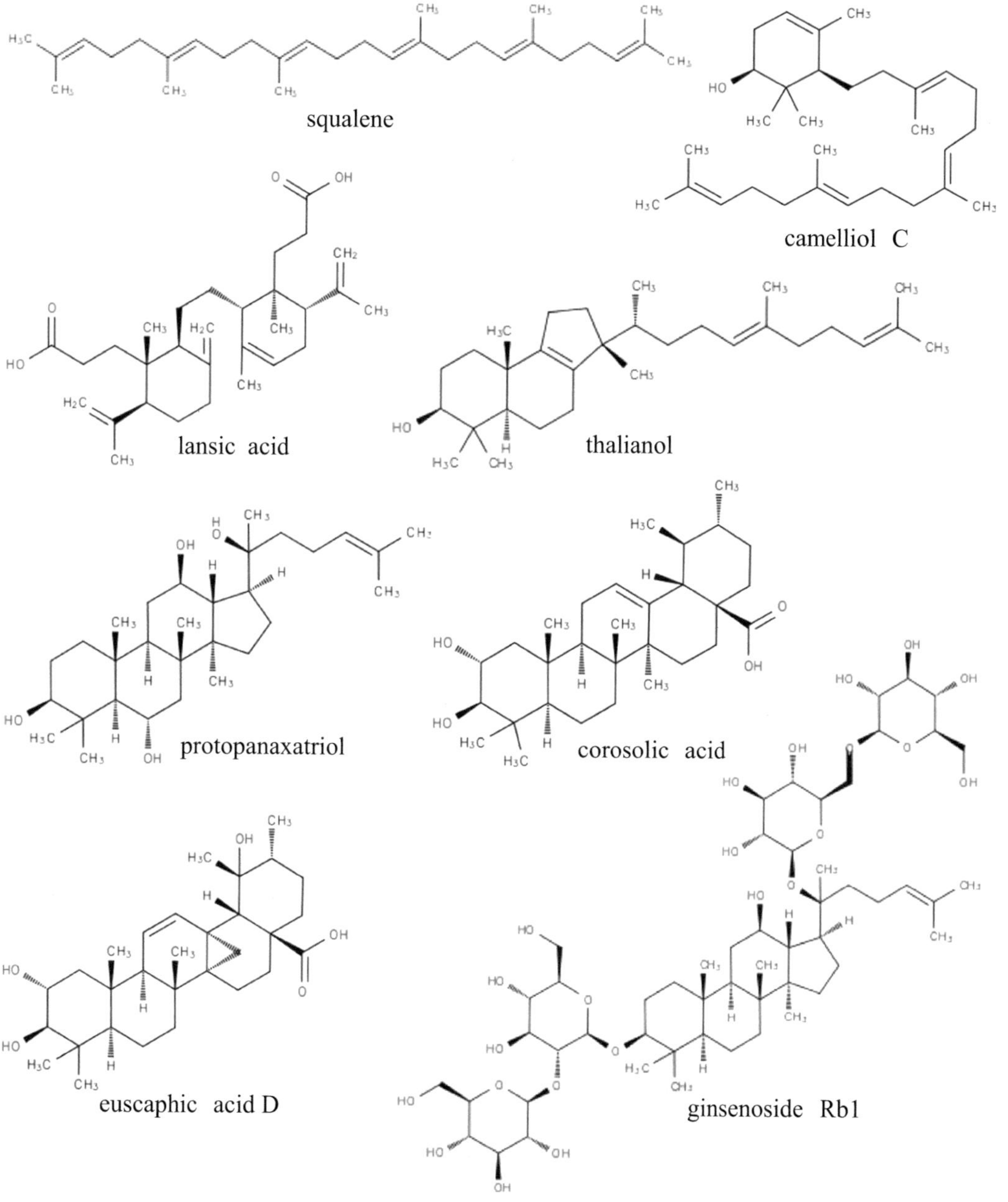

squalene

camelliol C

lansic acid

thalianol

protopanaxatriol

corosolic acid

euscaphic acid D

ginsenoside Rb1

Fig. 9.1 Different triterpene classes. Representatives of the acyclic (squalene), monocyclic (camelliol C), bicyclic (lansic acid), tricyclic (thalinol), tetracyclic (protopanaxatriol), pentacyclic (corosolic acid), hexacyclic (euscaphic acid D), and saponin (ginsenoside Rb1) are shown. Structures were obtained from the Chemical Entities of Biological Interest (ChEBI, http://www.ebi.ac.uk/chebi)

Saponins contain one or more hydrophilic glycoside moieties covalently linked with a lipophilic aglycone (i.e., sapogenin) that may bear triterpene or steroidal scaffold (Sawai and Saito 2011). Avenacin A-1, asiaticosides, gypsosaponins, ginsenosides, and soyasaponins found in *Avena sativa*, *Centella asiatica*, *Gypsophila oldhamiana*, *Panax* sp., and legume species, respectively, represent triterpene saponin class. However, asparasaponins, avenacosides, and digoxin biosynthesized in *Asparagus* sp, *Avena sativa,* and *Digitalis lanata*, respectively, are the steroidal saponins. Moreover, on the basis of hemolysis activity, saponins were also classified as hemolytic or nonhemolytic saponins (Biazzi et al. 2015). Hemolytic activity is dependent on the regio-specific oxidation of the sapogenin scaffolds. Oxidations at the C-23 and C-28 positions of sapogenin correlated with the hemolytic activity of saponins bearing hederagenin, zanhic acid, and medicagenic acid scaffolds. However, C-21, C-22, and C-24 oxidations of sapogenin were found in nonhemolytic saponins derived from soyasapogenols (Carelli et al. 2011; Tava et al. 2011).

9.3 Bioactive Triterpenes of *Ocimum*

Several triterpenes, mostly of the pentacyclic class, were isolated from *Ocimum* sp. and studied for their potential pharmacological activities (Fig. 9.2). The ursane and oleanane triterpenes ursolic acid and oleanolic acid, respectively, are the most abundant triterpenes of *Ocimum* sp. Ursolic acid and oleanolic acid were also considered as marker compounds for the development of many *Ocimum*-based standardized herbal medicines. Ursolic acid was detected in *O. basilicum, O. tenuiflorum, O. gratissimum, O. americanum, O. micranthum,* and *O. selloi* (Silva et al. 2008; Jager et al. 2009; Misra et al. 2014, 2017). The amount of ursolic acid in different *Ocimum* species varied considerably. Among *Ocimum* species, *O. tenuiflorum* accumulated higher level of ursolic acid in leaf (Silva et al. 2008). Oleanolic acid was isolated from *O. basilicum* and *O. tenuiflorum* (Misra et al. 2017). However, the presence of oleanolic acid in other *Ocimum* sp. cannot be excluded, considering the analytical constraints for the isomeric triterpenes in a complex plant extract. Ursolic acid and oleanolic acid possess diverse pharmacological activities such as hepatoprotective, anticarcinogenic, anti-inflammatory, antioxidant, and antimicrobial. Dietary supplements, over-the-counter drugs, and cosmetics containing ursolic acid and oleanolic acid as major ingredients are available in global markets. Besides, semisynthetic derivatives of oleanolic acid (bardoxolone methyl and omaveloxolone) are being tested in Phase 1 to 3 clinical studies for the remedial of medical complications such as chronic kidney diseases, pulmonary hypertension, mitochondrial myopathies, Friedreich's ataxia, melanoma (http://reatapharma.com/).

In addition to ursolic acid and oleanolic acid, several other bioactive triterpenes were also reported from hairy root cultures of *O. basilicum* developed by genetic transformation using *Agrobacterium rhizogenes* (Marzouk 2009). The triterpenes of the hairy root cultures were lupanes (betulinic and alphitolic acid), ursane (euscaphic acid) and oleanane (3-epi-maslinic). The production of betulinic acid in callus cultures of three *Ocimum* species (i.e., *O. basilicum, O. kilimandscharicum,* and *O. sanctum*) was also reported (Pandey et al. 2015). Considering the negligible amount of betulinic acid in *Ocimum* leaves, this work revealed the untapped potentials of *Ocimum* calli. Betulinic acid is well known for anticarcinogenic, anti-inflammatory, anti-HIV, and hepatoprotective activities (Lee et al. 2015; Zhang et al. 2015). Alphitolic acid, euscaphic acid, and 3-epi-maslinic also showed hepatoprotective activity (Marzouk 2009). In addition, alphitolic acid possesses anti-inflammatory and anticarcinogenic activities (Bai et al. 2015). Triterpene derivatives basilol, ocimol, and steroidal glycoside were also isolated from aerial parts of *O. basilicum*, although the bioactivities of these compounds remained to be tested (Siddiqui et al. 2007).

9.4 Spatiotemporal Accumulation of Triterpenes: Implication to the Biological Function

In contrast to the primary roles of sterols and steroid hormones, triterpenes are generally considered to have specialized functions in plants. For instance, avenacins (antimicrobial triterpene

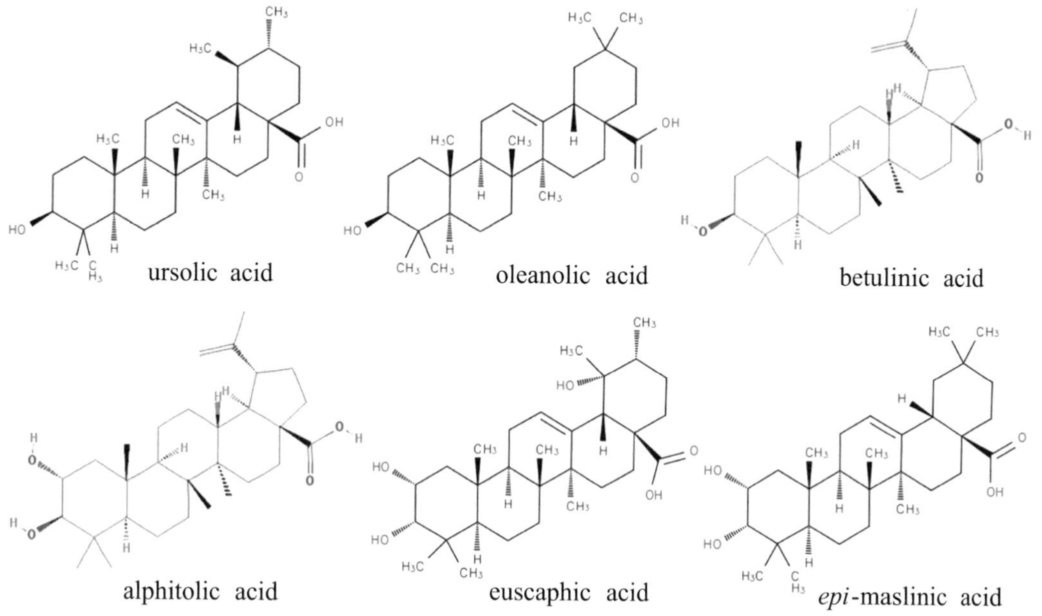

Fig. 9.2 Bioactive triterpenes of *Ocimum*. Structures were obtained from the Chemical Entities of Biological Interest (ChEBI, http://www.ebi.ac.uk/chebi) and Pub Chem (https://www.ncbi.nlm.nih.gov/pccompound)

saponins) are produced in roots of the oat plants to confer defense toward root-infecting fungal pathogens. Oat plants that could not produce avenacins in roots were susceptible to the root-infecting fungal pathogens such as *Gaeumannomyces graminis*, *Fusarium culmorum,* and *F. avenaceum* (Papadopoulou et al. 1999). Likewise, the catabolism of arabidiol (a tricyclic triterpene) was essential for the production of (E)-4,8-dimethyl-1,3,7-nonatriene, a volatile compound in Arabidopsis roots for defense towards the root rot pathogen *Pythium irregular* (Sohrabi et al. 2015). Tissue- and organ-preferential accumulation of several other triterpenes such as glycyrrhizin, ginsenosides, bacoside, ursolic acid, oleanolic acid, arjunetin was known in plants, including Glycyrrhiza, Panax, Bacopa, *Ocimum,* and Arjuna. However, the biological functions of most of the triterpenes were not elucidated (Ghosh 2016). Similarly, the genes of the triterpene biosynthetic pathway spatiotemporally express in plants. Besides, a few triterpenes such as β-amyrin, lupeol, thalianol, marneral-derived triterpenes, and poaceatapetol were suggested to perform/interfere with essential functions in plants during growth and developmental stages including nodulation, root development, flowering, embryogenesis, and reproduction (Delis et al. 2011; Go et al. 2012; Kemen et al. 2014; Xue et al. 2018).

In *O. basilicum*, spatiotemporal accumulation of ursolic acid and oleanolic acid, two major pentacyclic triterpenes of *O. basilicum,* was studied (Misra et al. 2014, 2017). Selective metabolite analysis of cuticular wax component of *O. basilicum* leaf and stem revealed a specific localization of ursolic acid and oleanolic acid in cuticle of the plant surface. The cuticle-specific accumulation of ursolic acid and oleanolic acid was correlated with the epidermal cell specialization of the biosynthetic pathway (Misra et al. 2014). In view of the protective roles of plant cuticle against pathogens and water loss, the localization of ursolic acid and oleanolic acid in cuticle might be linked with their potential function in plant defense toward biotic and abiotic stresses. Besides, the involvement of ursolic acid and oleanolic acid as nanofillers in providing the mechanical strength to the cuticular matrix was also suggested (Tsubaki et al. 2013).

In *O. basilicum*, the amount of ursolic acid and oleanolic acid in different organs varied considerably (Misra et al. 2014). The highest levels of ursolic acid and oleanolic acid were detected in inflorescences followed by leaves, stems, and roots. Interestingly, among the differently-aged leaves, increased amount of ursolic acid and oleanolic acid was noticed in the youngest and older leaves, respectively. Ursolic acid and oleanolic acid biosynthetic pathway genes also exhibited organ-preferential expression in *O. basilicum* (Misra et al. 2017). Organ-preferential accumulation of ursolic acid and oleanolic acid indicated their physiological function in *O. basilicum* in an organ-specific manner. Preferential accumulation in reproductive organ and antimicrobial activity of ursolic acid and oleanolic acid suggested a role in providing protection against microbial attacks. Similar to the role of poaceatapetol in rice, ursolic acid and oleanolic acid might also prevent desiccation of pollens during reproduction in *O. basilicum* (Xue et al. 2018). The treatment of *O. basilicum* with methyl jasmonate (MeJA), a phytohormone of the plant defense pathway, resulted in increased production of ursolic acid and oleanolic acid in leaf. Moreover, ursolic acid and oleanolic acid biosynthetic pathway genes were also overexpressed in response to infection with phytopathogen *Sclerotinia sclerotiorum* and treatments with MeJA and ethylene (a phytohormone functioning in plant defense), suggesting the role of ursolic acid and oleanolic acid in plant defense (Misra et al. 2014, 2017). To determine the actual biological function to ursolic acid and oleanolic acid, gene overexpression or silencing studies in *O. basilicum* appear to be crucial.

9.5 An Overview of Triterpene Pathway

All the structurally diverse classes of terpenes are biosynthesized from two 5-carbon isoprene units, isopentenyl diphosphate (IPP), and dimethylallyl diphosphate (DMAPP) that can be generated from either of two cellular pathways, the mevalonic acid (MVA) pathway and methylerythritol phosphate (MEP) pathway, involving seven and eight enzymatic reactions, respectively. In fungi and animals, MVA pathway operates for IPP and DMAPP generation from acetyl-CoA. In prokaryotes, MEP pathway generates IPP and DMAPP using pyruvic acid and glyceraldehyde-3-phosphate. In higher plants, both MEV and MEP pathways are active in different subcellular compartments to produce IPP and DMAPP pool in cells. In plants, IPP and DMAPP derived from the cytoplasmic MVA pathway are channelized for the formation of sesquiterpenes (C15), triterpenes (C30), and polyprenols (>45). However, plastidial MEP pathway supplies IPP and DMAPP for the production of monoterpenes (C10), diterpenes (C20), sesterterpenes (C25), carotenoids (C40), and long-chain phytol.

In cytoplasm, C15 farnesyl pyrophosphate (FPP), the precursor for sesquiterpenes, is produced by the enzyme FPP synthase following coupling of two units of IPP and one unit of DMAPP. In the triterpene pathway, squalene synthase catalyzes the formation of C30 squalene using two units of FPP. Further, squalene epoxidation by squalene epoxidase leads to 2,3-oxidosqualene which is subsequently converted to diverse steroidal and triterpene scaffolds in plants (Fig. 9.3). 2,3-Oxidosqualene cyclization by oxidosqualene cyclase (OSC) marks the initial diversifying step in the biosynthesis of diverse triterpene and sterol scaffolds in plants. Cycloartenol synthase (CAS) cyclizes 2,3-oxidosqualene to cycloartenol from which membrane sterols and steroid hormones are produced in plants (Thimmappa et al. 2014; Ghosh 2016). Besides CAS for the generation of sterol scaffold, a number of OSCs operate in plants for the formation of diverse triterpene scaffolds, ranging from one to five rings structures. For example, a total of thirteen and nine OSCs are known in *Arabidopsis thaliana* and *Oryza sativa*, respectively, including one CAS in each species. On the contrary, in animals and fungi, a single OSC (lanosterol synthase) is active for the formation of membrane sterols and steroid hormones. Subsequent to OSC-mediated cyclization, triterpene scaffolds undergo diverse

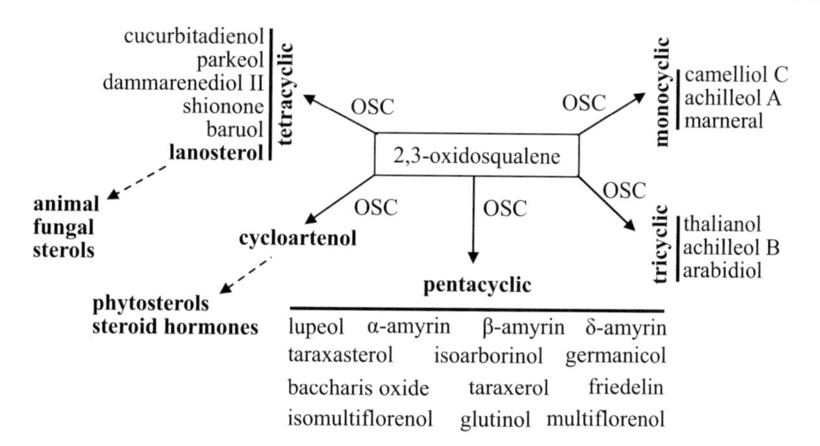

Fig. 9.3 The origin of diverse cyclic triterpene scaffolds from 2,3-oxidosqualene, the general triterpene precursor. Oxidosqualene cyclases (OSC) catalyze the conversion of 2,3-oxidosqualene into cyclic triterpene scaffolds

type of regio- and sterio-specific modifications including oxidation and glycosylation catalyzed by P450s and UGTs (Osbourn et al. 2011; Ghosh 2017).

9.6 Transcriptomic and Genomic Studies in *Ocimum*

The biosynthesis of triterpenes in *Ocimum* sp. had been known for a long time (Nicholas 1961). However, understanding the triterpene biosynthetic pathway and identification and characterization of the pathway enzymes/genes are relatively recent events, owing to the recent advancement in the high-throughput transcriptomic and genomic tools (Misra et al. 2014, 2017). Following sections discuss transcriptomic and genomic studies that were useful for the identification of genes/enzymes of the triterpene biosynthetic pathway in *Ocimum* species.

9.6.1 MeJA-Responsive Transcripts

The amount of triterpenes (ursolic acid and oleanolic acid) in the leaves of *O. basilicum* was increased following MeJA treatment (Misra et al. 2014). This enhancement in triterpene biosynthesis was hypothesized to be due to the increased rate of transcription of the biosynthetic pathway genes in response to MeJA elicitation. To selectively identify mRNAs that are expressed in *O. basilicum* in response to MeJA elicitation, suppression subtractive hybridization (SSH) was carried out (Misra et al. 2014). In SSH approach, two populations of mRNAs (MeJA-treated sample vs. untreated control sample) were compared to obtain the cDNA clones of the genes that were expressed in high level following MeJA elicitation in *O. basilicum.* This approach identified a total of 388 MeJA-responsive unique transcripts of *O. basilicum.* MeJA-responsive transcripts represented genes of the MeJA biosynthetic pathway, plant defense pathways, several transcriptional regulators, and diverse secondary metabolic pathways such as terpenes and phenylpropanes. Thus, MeJA controlled its own biosynthesis as well as several defense and secondary metabolic pathways of *O. basilicum.* A number of transcripts were originated from the triterpene pathway genes, including mevalonate kinase, farnesyl diphosphate synthase, oxidosqualene cyclase, and P450. Mevalonate kinase is an enzyme of the MEV pathway that supplies 5C IPP and DMAPP for the triterpene biosynthesis. Farnesyl diphosphate synthase produces 15C FPP, the precursor for 30C triterpenes. Oxidosqualene cyclase and P450 catalyze triterpene cyclization and oxidative reactions, respectively. Moreover, efforts have also been made to identify transcripts for the *O. basilicum* genes that are expressed in glandular trichomes, i.e., specialized glands on the plant surface (Gang et al. 2001; Kapteyn et al. 2007). In *Ocimum* sp., secretory glandular trichomes are the site for the biosynthesis and storage of essential oil components including phenylpropanes, eugenol, and methyl chavicol that are biosynthesized from the amino acid phenylalanine. About 30,000 expressed sequence tags (ESTs) were identified that represented glandular trichome-expressed genes of the volatile phenylpropane biosynthetic pathway. Glandular trichome-expressed ESTs were assembled into a total of about 6000 unique transcripts, lacking any transcripts of the triterpene biosynthesis pathway genes such as oxidosqualene cyclases and P450s. The epidermal cell specialization of the triterpene biosynthesis pathway might have caused negligible expression of the triterpene biosynthesis pathway genes in *O. basilicum* glandular trichomes (Misra et al. 2014).

9.6.2 High-Throughput Transcriptome and Genome Sequencing

In recent times, the availability of low-cost and high-throughput sequencing technologies revolutionized genome-wide analysis of the gene expression and identification of the novel genes (Loman et al. 2012). Using Illumina high-throughput sequencing platform, the nuclear and chloroplast genomes of *O. tenuiflorum* (*O. sanctum*) were studied (Rastogi et al. 2015; Upadhyay et al. 2015). The nuclear genome of *O. tenuiflorum* revealed candidate genes of the specialized metabolite pathways including phenylpropane and terpene biosynthetic pathways. Among the triterpene biosynthetic pathway genes, potential oxidosqualene cyclases and P450s were identified in the *O. tenuiflorum* genome. A comparative analysis of the transcriptomes of *O. basilicum* and *O. sanctum* was also carried out (Rastogi et al. 2014). These two *Ocimum* species displayed contrasting essential oil composition and differed in the expression profiles of the genes of the phenylpropane and terpene biosynthetic pathways. Moreover, transcriptomic studies were also conducted in *O. americanum* and *O. basilicum* to understand cold stress response and the role of anthocyanins in photoprotection, respectively (Zhan et al. 2016; Torre et al. 2016). Collectively, high-throughput genome and transcriptome sequencing in *Ocimum* species resulted in a huge amount of data related to the gene sequences and expression profiles of the genes. However, functional genomics studies on the selected biochemical pathways and genes need to be conducted to understand the genetic basis for distinct metabolite profiles in diverse *Ocimum* species.

9.7 Available Tools for Gene Function Analysis in *Ocimum*

The sequencing of *Ocimum* genome led to the identification of thousands of novel genes whose functions are yet to be assigned. Nevertheless, high-throughput transcriptome sequencing revealed putative function of the genes based on differential transcript expression patterns in *Ocimum* species. However, a detailed functional analysis of the individual genes is indispensable to establish the actual biological function of the *Ocimum* genes. The functional role of the genes can be determined by conducting gene overexpression or silencing studies. Following sections discuss the currently available tools for conducting gene overexpression and silencing studies in *Ocimum*.

9.7.1 Virus-Induced Gene Silencing

In response to attack by viruses, plants activate an efficient RNA-mediated antiviral defense pathway that targets the viral genome for the sequence homology-dependent silencing (Ruiz et al. 1998). In this process, double-stranded viral

RNAs produced by the viral genome-encoded RNA-dependent RNA polymerase (RdRP) is recognized by the plants as aberrant RNAs and is processed by DICER, an endonuclease, into small interfering RNAs (siRNAs). siRNAs guide an RNA-induced silencing complex (RISC), including an endonuclease argonaute, for the homology-dependent cleavage of the target RNAs. This antiviral defense mechanism was exploited to develop virus-induced gene silencing (VIGS) tool for gene function analysis in plants (Liu et al. 2002; Lu et al. 2003). Several VIGS vectors were developed based on either RNA or DNA viruses such as tobacco rattle virus, potato virus X, tobacco mosaic virus, barley stripe mosaic virus, tomato golden mosaic virus (Robertson 2004). In order to silence a host gene, the sequence of the target gene is inserted into the VIGS vector, and the vector is delivered into the host cells to trigger siRNA-mediated gene silencing pathway. Among the different VIGS vectors, tobacco rattle virus (TRV)-based vectors are quite popular for wide host range and were useful for conducting VIGS in several plant species, including tobacco, tomato, arabidopsis, *Ocimum*, Madagascar periwinkle, ashwagandha, cotton.

The application of VIGS technology in *Ocimum* is a recent development (Misra et al. 2017). TRV-based VIGS has been demonstrated in *O. basilicum* by tracking the silencing of the VIGS marker gene, protoporphyrin IX magnesium chelatase subunit H (ChlH). The silencing of ChlH resulted in a readily visible leaf-yellowing phenotype due to the loss in chlorophyll biosynthesis (Fig. 9.4a). The leaf-yellowing phenotype, indication of the *ChlH* silencing, was regularly noticed with 73% frequency in the first leaf pair to emerge after inoculation with the VIGS vectors. VIGS resulted in c. 63% suppression of the transcript level of *ChlH* and c. 50% loss in chlorophyll content in *O. basilicum* leaves (Fig. 9.4b). By employing TRV-based VIGS tool, the involvement of P450s in *in planta* biosynthesis of bioactive triterpenes ursolic acid and oleanolic acid in *O. basilicum* was determined (Misra et al. 2017).

9.7.2 Transient RNA Interference and Gene Overexpression

Similar to VIGS, gene silencing through RNA interference (RNAi) is dependent on the accumulation of double-stranded RNAs in cells that lead to the generation of siRNAs by DICER which triggers sequence-based RISC-mediated degradation of the target mRNAs (Fellmann and Lowe 2014). Unlike the involvement of RdRP for double-stranded RNA formation in VIGS, RNA expressed from the RNAi vectors is folded into hairpin structure which is processed into siRNAs. In recent times, transient RNAi and gene overexpression studies have been conducted in *Ocimum* to understand gene function. Transient RNAi was demonstrated in *O. sanctum* using pHANNIBAL/pART27 and pRI101-AN vectors (Rastogi et al. 2013; Jayaramaiah et al. 2016). The expression of hairpin RNAs specific to 4-coumarate: CoA ligase (4CL) resulted in c. 40% reduction in gene transcript level in *O. sanctum*. Based on the transient RNAi approach, the involvement of a 4CL in *in planta* biosynthesis of eugenol was determined (Rastogi et al. 2013). By employing transient RNAi and gene overexpression, Jayaramaiah et al. 2016, elucidated the role of a sesquiterpene synthase (*OkBCS*) in the accumulation of β-caryophyllene in *O. kilimandscharicum*. Up to 95% reduction in *OkBCS* transcripts was achieved in RNAi experiment. However, gene overexpression experiment led to 400% higher expression of *OkBCS* transcripts.

9.7.3 Hairy Root Culture

Hairy roots are produced in plants in response to infection by the gram-negative soil bacterium *A. rhizogenes*. The bacterium carries a transfer DNA (T-DNA) in the root-inducing (Ri) plasmid that encodes root locus (rol) genes *rolA*, *rolB*, and *rolC*. During plant infection, *A. rhizogenes* transforms its T-DNA into plant genome and thus, triggers the formation of hairy roots by changing the levels of phytohormones (auxin and

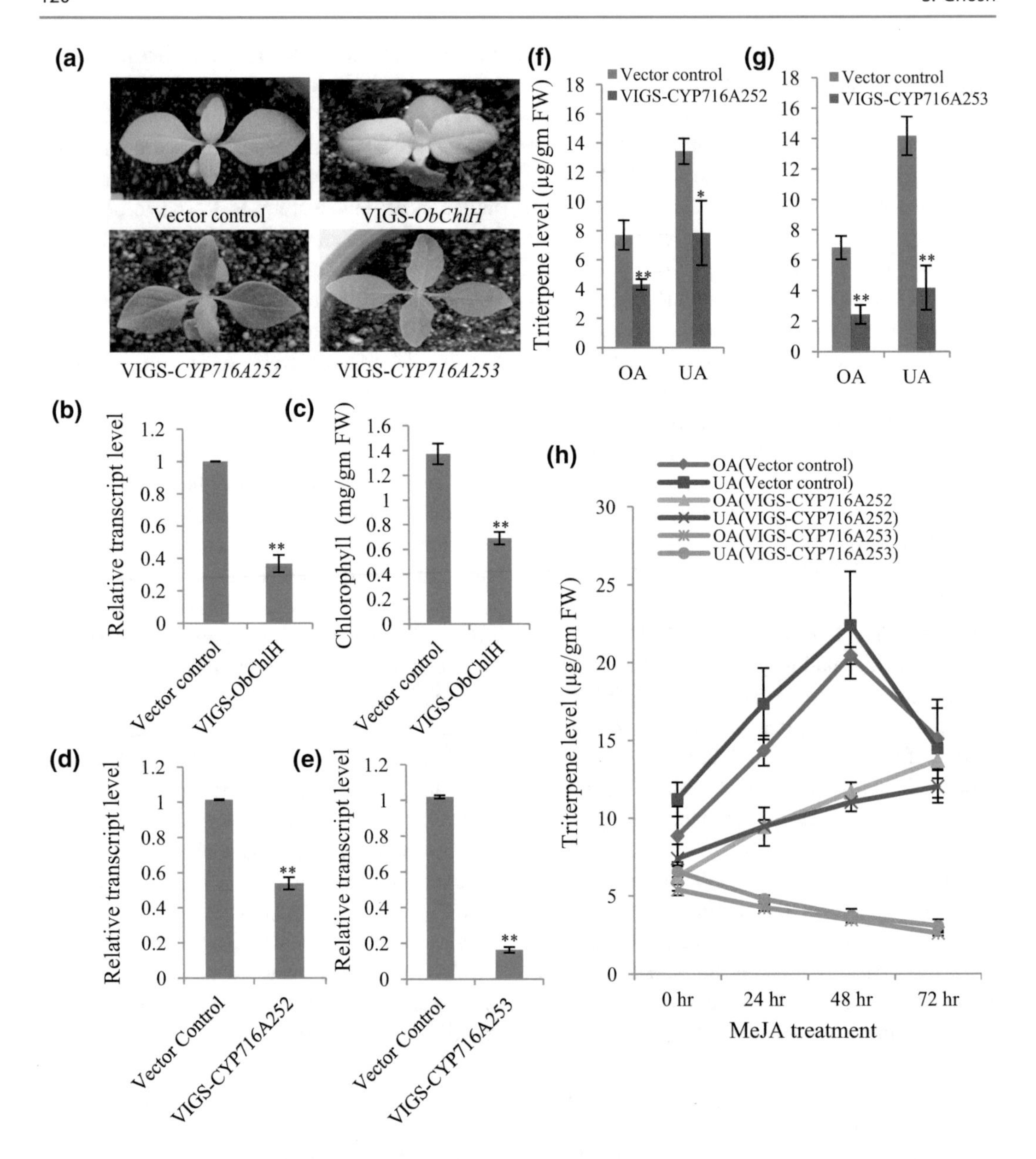

Fig. 9.4 Virus-induced gene silencing in *O. basilicum*. **a** Virus-induced gene silencing (VIGS) of *ObchlH* led to leaf-yellowing phenotype (marked by arrows). **b** Relative transcript level of *ObchlH* in vector control and *VIGS-ObChlH* seedling was determined by quantitative reverse transcription—polymerase chain reaction (qRT-PCR). **c** Total chlorophyll amount (Chla+b) in vector control and *VIGS-ObChlH* seedlings. FW denotes fresh weight. **d** Relative transcript level of *CYP716A252* in vector control and *VIGS-CYP716A253* seedlings. **e** Relative transcript level of *CYP716A253* in vector control and *VIGS-CYP716A253* seedlings. **f**, **g** Oleanolic acid (OA) and ursolic acid (UA) content in vector control and **f** VIGS-C*YP716A252* or **g** VIGS-*CYP716A253* seedlings. **h** OA and UA content in vector control, VIGS-CYP716A252, and VIGS-*CYP716A253* seedlings after methyl jasmonate (MeJA) treatment. Figures are reproduced, with permission, from Misra et al. (2017)

cytokinin) in the cells. Hairy root lines are genetically stable and display hormone-independent fast growth in culture. Hairy root cultures obtained from diverse plant species showed the ability to carry out biotransformation of pharmaceutically relevant compounds (Banerjee et al. 2012). Hairy root technology was applied to study the function of gene promoters in tomato (*Solanum lycopersicum*), *S. pennellii,* and *Lepidium hyssopifolium* (Ron et al. 2014). Ron et al. demonstrated that targeted gene mutation can be introduced into the plant genome by employing CRISPR/Cas9 system in tomato through hairy root transformation. Several research groups reported the development of hairy root lines in *O. basilicum* for the production of valuable secondary metabolites including triterpenes, rosmarinic acid, and antioxidants (Tada et al. 1996; Marzouk 2009; Srivastava et al. 2016). Thus, hairy root technology can be employed in *Ocimum* species for studying gene function.

9.7.4 Stable Transgenics

Agrobacterium tumefaciens-mediated gene transfer technology has revolutionized the research on gene function and also led to the generation of stable transgenic plants with improved agronomic and quality traits (Gelvin 2003; Datta and Ghosh 2015). *A. tumefaciens*, a crown gall disease-causing gram-negative soil bacterium, carries a T-DNA in the tumor-inducing (Ti) plasmid that is integrated into the plant genome during the infection process. A number of disarmed binary vectors are developed by engineering of Ti plasmid (Lee and Gelvin 2008). These binary vectors are suitable to introduce desire DNA sequence into the plant genome. The generation of stable transgenic lines following *A. tumefaciens*-mediated transformation has been demonstrated in *Ocimum* species such as *O. basilicum* (Deschamps and Simon 2002; Wang et al. 2016), *O. gratissimum* (Khan et al. 2015), and *O. citriodorum* (Deschamps and Simon 2002). The regulatory roles of spearmint (*Mentha spicata*) MYB and YABBY5 transcription factors in the terpene biosynthesis have been determined by conducting gene overexpression studies in transgenic *O. basilicum* (Wang et al. 2016; Reddy et al. 2017). Thus, transgenic technology can be applied to *Ocimum* species for the gene function analysis.

9.8 Functional Studies on the Key Enzymes of the *Ocimum* Triterpene Pathway

The enzymes and the corresponding genes of *O. basilicum* for the biosynthesis of bioactive pentacyclic triterpenes (ursolic acid and oleanolic acid) were recently identified and functionally characterized (Misra et al. 2014, 2017). These key enzymes included oxidosqualene cyclases (OSCs) and P450s of the triterpene biosynthetic pathway. However, the complete sets of triterpene biosynthetic pathway enzymes of the diverse *Ocimum* species are yet to be identified and functionally characterized. In the following section, the functional roles of OSCs and P450s in the biosynthesis of ursolic acid and oleanolic acid in *O. basilicum* are discussed.

9.8.1 Oxidosqualene Cyclases

OSCs catalyze the first diversifying step in triterpene pathway by converting 2,3-oxidosqualene into a variety cyclic triterpene scaffolds (Fig. 9.3; Abe 2014; Ghosh 2016). *O. basilicum* accumulated higher amounts of ursolic acid and oleanolic acid in response to treatment with MeJA (Misra et al. 2014). Two OSCs (ObAS1 and ObAS2) of *O. basilicum* were identified by screening a collection of ESTs, representing genes expressed in response to MeJA treatment and by performing polymerase chain reaction (PCR) using degenerate oligonucleotides (Misra et al. 2014). ObAS1 and ObAS2 encode for 761- and 765-amino acid proteins, respectively, that contain amino acid motifs characteristic of the OSC family members. These amino acid motifs are a DCTAE motif located in the enzyme active site for 2,3-oxidosqualene binding and protonation, a MWCYCR motif and a few QW motifs. MWCYCR motif might function in determining product profiles of OSCs (Kushiro et al. 2000). Whereas, QW motifs located at the surface of the enzyme might take part in stabilizing the enzyme during highly exothermic triterpene cyclization reaction (Wendt et al. 1997).

ObAS1 was functionally characterized as a β-amyrin synthase that cyclized 2,3-oxidosqualene to β-amyrin. In contrast, ObAS2 was found to be an unusual mixed amyrin synthase that catalyzed the formation of both α-amyrin and β-amyrin using 2,3-oxidosqualene substrate. Nevertheless, ObAS2 showed a preference for α-amyrin product as compared to β-amyrin. α-Amyrin and β-amyrin are the direct precursors for the biosynthesis of oleanolic acid and ursolic acid, respectively. Thus, ObAS1 and ObAS2 catalyzed the penultimate reactions in the ursolic acid and oleanolic acid biosynthetic pathway (Fig. 9.5). ObAS1 and ObAS2 showed 60% identity at the amino acid sequence level and contained all the conserved amino acid motifs (DCTAE, MWCYCR and QW). Therefore,

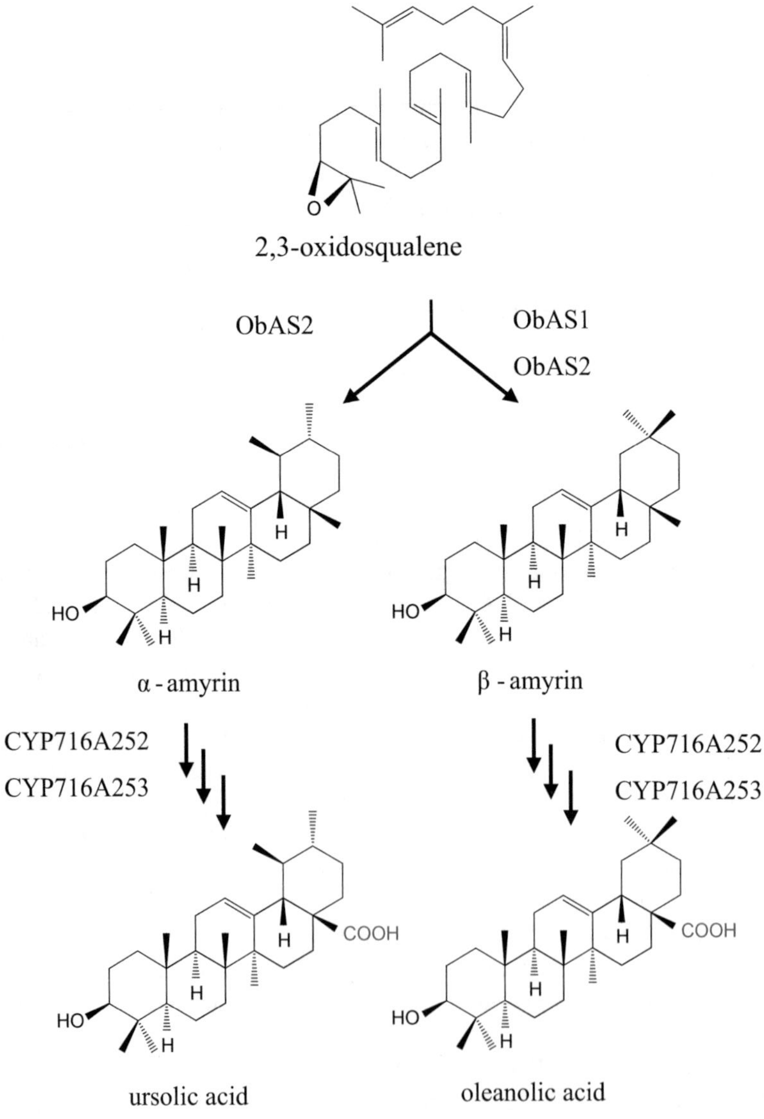

Fig. 9.5 Ursolic acid and oleanolic acid biosynthetic pathway in *O. basilicum*. ObAS1, β-amyrin synthase, ObAS2, mixed amyrin synthase having product preference to α-amyrin, CYP716A252 and CYP716A253, α-/β-amyrin C-28 oxidases. Multiple arrows denote three-step oxidation of α-/β-amyrin to ursolic acid/oleanolic acid

Misra et al. (2014) suggested that the variation in product profiles of ObAS1 and ObAS2 could be because of the difference in amino acid residues that are located outside the conserved motifs for the OSC family members. Although, a comparison of ObAS1 and ObAS2 revealed amino acid residues potentially being involved in differential product profiles, further analysis following swapping/site-directed mutagenesis approaches are required to confirm their role.

The preferential expression of ObAS1 and ObAS2 in the epidermal tissues indicated the epidermal cell specialization for the ursolic acid and oleanolic acid biosynthetic pathway. However, ursolic acid and oleanolic acid accumulate in the cuticle of the *O. basilicum*. These observations indicated the difference in the sites for biosynthesis and storage of ursolic acid and oleanolic acid, and the role of the transporter for their secretion into the cuticle.

9.8.2 P450s of the CYP716A Subfamily

The oxidation of α-amyrin and β-amyrin at the C28 position is the final reaction in the ursolic acid and oleanolic acid biosynthetic pathway (Fig. 9.5). P450s potentially involved in the C28 oxidation of α-/β-amyrin scaffolds were identified from the *O. basilicum* EST collections developed using MeJA-treated plant samples (Misra et al. 2014). Two P450s of the CYP716A subfamily (CYP716A252 and CYP716A253) were found to catalyze C28 oxidation of α-amyrin and β-amyrin leading to ursolic acid and oleanolic acid, respectively (Fig. 9.5; Misra et al. 2017). CYP716A252 and CYP716A253 encode for polypeptides of 478 and 477 amino acids, respectively. CYP716A252 and CYP716A253 contain amino acid motifs (FxxGxRxCxG, ERR triad, A/GGXD/ET) characteristics of bona fide P450s. Combinatorial expression of *O. basilicum* amyrin synthase (ObAS1/ObAS2) and amyrin C28 oxidase (CYP716A252/CYP716A253) along with a cytochrome P450 reductase (AtCPR1) resulted in the formation of ursolic acid and oleanolic acid in yeast (*Saccharomyces cerevisiae*). The titer of ursolic acid and oleanolic acid in the yeast shake flask culture was c. 0.5 and 3 mg/l, respectively. Interestingly, CYP716A253 was more efficient than CYP716A252 for the α-/β-amyrin C28 oxidation in yeast (Misra et al. 2017). Two cytochrome P450 reductases (ObCPR1 and ObCPR2) that can function as redox partners of the CYP716A252 and CYP716A253 were also identified in *O. basilicum* (Misra et al. 2017). Altogether these studies revealed the utility of *O. basilicum* amyrin synthases, amyrin C28 oxidases, and cytochrome P450 reductases for the production of bioactive triterpenes in heterologous host yeast.

To clarify the roles of CYP716A252 and CYP716A253 in the *in planta* biosynthesis of the ursolic acid and oleanolic acid, VIGS was employed in *O. basilicum* (Misra et al. 2017). Transcript expression of *CYP716A252* and *CYP716A253* was reduced up to c. 50 and 80%, respectively, in VIGS seedlings (Fig. 9.4d, e). *CYP716A252* and *CYP716A253* suppressions also resulted in reduction in the amount of ursolic acid and oleanolic acid in the leaves (Fig. 9.4f, g). These results suggested that CYP716A252 and CYP716A253 are involved in the biosynthesis of ursolic acid and oleanolic acid in the leaves of *O. basilicum*. However, CYP716A253 played a major role in MeJA-induced accumulation of ursolic acid and oleanolic acid in *O. basilicum* (Fig. 9.4h). The nucleotide and amino acid sequences of CYP716A252 and CYP716A253 were 78 and 85% identical, respectively. Misra et al. 2017 proposed that two amyrin C28 oxidases in *O. basilicum* is the result of a gene duplication event. After gene duplication event, amyrin C28 oxidases retained same biochemical function, however, evolved at the transcriptional control to participate in the constitutive and elicitor-mediated biosynthesis of ursolic acid and oleanolic acid in *O. basilicum*.

9.9 Concluding Remarks

Bioactive triterpenes are known to produce in a number of *Ocimum* species including *O. basilicum, O. tenuiflorum, O. gratissimum, O. americanum, O. micranthum, O. kilimandscharicum,* and *O. selloi*. However, the genes/enzymes for the biosynthesis of bioactive triterpenes were only studied in *O. basilicum*. Functional genomics studies conducted in *O. basilicum* led to the identification and functional characterization of OSCs and P450s that converted 2,3-oxidosqualene, the general triterpene precursors, to bioactive pentacyclic triterpenes ursolic acid and oleanolic acid. Recent advances in high-throughput sequencing of genomes and transcriptomes, and establishment of the functional genomics tools such as transgenics, RNA interference (RNAi), virus-induced gene silencing (VIGS), and hairy root culture will be instrumental in understanding triterpene biosynthesis in diverse *Ocimum* species.

References

Abe I (2014) The oxidosqualene cyclases: one substrate, diverse products. In: Osbourn A, Goss RJ, Carter GT (eds) Natural products: discourse, diversity, and design. Wiley, Hoboken, NJ, USA, pp 293–316

Arendt P, Miettinen K, Pollier J, De Rycke R, Callewaert N, Goossens A (2017) An endoplasmic reticulum-engineered yeast platform for overproduction of triterpenoids. Metab Eng 40:165–175

Bai LY, Chiu CF, Chiu SJ, Chen YW, Hu JL, Wu CY, Weng JR (2015) Alphitolic acid, an anti-inflammatory triterpene, induces apoptosis and autophagy in oral squamous cell carcinoma cells, in part, through a p53-dependent pathway. J Funct Foods 18:368–378

Banerjee S, Singh S, Ur Rahman L (2012) Biotransformation studies using hairy root cultures - a review. Biotechnol Adv 30:461–468

Biazzi E, Carelli M, Tava A, Abbruscato P, Losini I, Avato P, Scotti C, Calderini O (2015) CYP72A67 catalyzes a key oxidative step in *Medicago truncatula* haemolytic saponin biosynthesis. Mol Plant 8:1493–1506

Carelli M, Biazzi E, Panara F, Tava A, Scaramelli L, Porceddu A, Graham N, Odoardi M, Piano E, Arcioni S, May S, Scotti C, Calderini O (2011) *Medicago truncatula* CYP716A12 is a multifunctional oxidase involved in the biosynthesis of hemolytic saponins. Plant Cell 23:3070–3081

Datta A, Ghosh S (2015) Agricultural biotechnology for food sufficiency and benefit to human health (Chap. 72). In: Talwar GP, Hasnain SE, Sarin SK (eds) Textbook of biochemistry, biotechnology, allied & molecular medicine, under section VIII, human genetics, biochemical basis of inheritance, expression of genetic information, genetic engineering. Prentics Hall, India, pp 877–896

Delis C, Krokida A, Georgiou S, Pena-Rodriguez LM, Kavroulakis N, Ioannou E, Roussis V, Osbourn AE, Papadopoulou KK (2011) Role of lupeol synthase in *Lotus japonicus* nodule formation. New Phytol 189:335–346

Deschamps C, Simon JE (2002) *Agrobacterium tumefaciens*-mediated transformation of *Ocimum basilicum* and *O. citriodorum*. Plant Cell Rep 21:359–364

Fellmann C, Lowe SW (2014) Stable RNA interference rules for silencing. Nat Cell Biol 16(1):10–18

Gang DR, Wang J, Dudareva N, Nam KH, Simon JE, Lewinsohn E, Pichersky E (2001) An investigation of the storage and biosynthesis of phenylpropenes in sweet basil. Plant Physiol 125:539–555

Gelvin SB (2003) *Agrobacterium*-mediated plant transformation: the biology behind the “gene-jockeying” tool. Microbiol Mol Biol Rev 67:16–37

Ghosh S (2016) Biosynthesis of structurally diverse triterpenes in plants: the role of oxidosqualene cyclases. Proc Indian Nat Sci Acad 82:1189–1210

Ghosh S (2017) Triterpene structural diversification by plant cytochrome P450 enzymes. Front Plant Sci 8:1886

Go YS, Lee SB, Kim HJ, Kim J, Park HY, Kim JK (2012) Identification of marneral synthase, which is critical for growth and development in Arabidopsis. Plant J 72:791–804

Hill RA, Connolly JD (2013) Triterpenoids. Nat Prod Rep 30:1028–1065

Hill RA, Connolly JD (2017) Triterpenoids. Nat Prod Rep 34:90–122

Jager S, Trojan H, Kopp T, Laszczyk MN, Scheffler A (2009) Pentacyclic triterpene distribution in various plants-rich sources for a new group of multipotent plant extracts. Molecules 14:2016–2031

Jayaramaiah RH, Anand A, Beedkar SD, Dholakia BB, Punekar SA, Kalunke RM, Gade WN, Thulasiram HV, Giri AP (2016) Functional characterization and transient expression manipulation of a new sesquiterpene synthase involved in β-caryophyllene accumulation in *Ocimum*. Biochem Biophys Res Commun 473:265–271

Kapteyn J, Qualley AV, Xie Z, Fridman E, Dudareva N, Gang DR (2007) Evolution of cinnamate/p-coumarate carboxyl methyltransferases and their role in the biosynthesis of methylcinnamate. Plant Cell 19:3212–3229

Kemen AC, Honkanen S, Melton RE, Findlay KC, Mugford ST, Hayashi K, Haralampidis K, Rosser SJ, Osbourn A (2014) Investigation of triterpene synthesis and regulation in oats reveals a role for β-amyrin in determining root epidermal cell patterning. Proc Natl Acad Sci USA 111:8679–8684

Khan S, Fahim N, Singh P, Rahman LU (2015) *Agrobacterium tumefaciens* mediated genetic transformation of *Ocimum gratissimum*: a medicinally important crop. Indust Crops Prod 71:138–146

Kushiro T, Shibuya M, Masuda K, Ebizuka Y (2000) Mutational studies on triterpene synthases: engineering lupeol synthase into β-amyrin synthase. J Am Chem Soc 122:6816–6824

Lee LY, Gelvin SB (2008) T-DNA binary vectors and systems. Plant Physiol 146:325–332

Lee SY, Kim HH, Park SU (2015) Recent studies on betulinic acid and its biological and pharmacological activity. EXCLI J 14:199–203

Li YC, Tian K, Sun LJ, Long H, Li LJ, Wu ZZ (2016) A new hexacyclic triterpene acid from the roots of *Euscaphis japonica* and its inhibitory activity on triglyceride accumulation. Fitoterapia 109:261–265

Liang CQ, Shi YM, Li XY, Luo RH, Li Y, Zheng YT, Zhang HB, Xiao WL, Sun HD (2013) Kadcotriones AC: tricyclic triterpenoids from *Kadsura coccinea*. J Nat Prod 76:2350–2354

Liu Y, Schiff M, Marathe R, Dinesh-Kumar SP (2002) Tobacco Rar1, EDS1 and NPR1/NIM1 like genes are required for N-mediated resistance to tobacco mosaic virus. Plant J 30:415–429

Loman NJ, Misra RV, Dallman TJ, Constantinidou C, Gharbia SE, Wain J, Pallen MJ (2012) Performance comparison of benchtop high-throughput sequencing platforms. Nat Biotechnol 30:434–439

Lu R, Martin-Hernandez AM, Peart JR, Malcuit I, Baulcombe DC (2003) Virus-induced gene silencing in plants. Methods 30:296–303

Marzouk AM (2009) Hepatoprotective triterpenes from hairy root cultures of *Ocimum basilicum* L. Z Naturforsch C 64:201–209

Meng FY, Sun JX, Li X, Yu HY, Li SM, Ruan HL (2011) Schiglautone A, a new tricyclic triterpenoid with a unique 6/7/9-fused skeleton from the stems of *Schisandra glaucescens*. Org Lett 13:1502–1505

Misra RC, Maiti P, Chanotiya CS, Shanker K, Ghosh S (2014) Methyl jasmonate-elicited transcriptional responses and pentacyclic triterpene biosynthesis in sweet basil. Plant Physiol 164:1028–1044

Misra RC, Sharma S, Sandeep Garg A, Chanotiya CS, Ghosh S (2017) Two CYP716A subfamily cytochrome P450 monooxygenases of sweet basil play similar but nonredundant roles in ursane- and oleanane-type pentacyclic triterpene biosynthesis. New Phytol 214:706–720

Moses T, Pollier J, Thevelein JM, Goossens A (2013) Bioengineering of plant (tri)terpenoids: from metabolic engineering of plants to synthetic biology in vivo and in vitro. New Phytol 200:27–43

Nicholas HJ (1961) Determination of sterol and triterpene content of *Ocimum basilicum* and *Salvia officinalis* at various stages of growth. J Pharm Sci 50:645–647

Osbourn A, Goss RJ, Field RA (2011) The saponins: polar isoprenoids with important and diverse biological activities. Nat Prod Rep 28:1261–1268

Pandey H, Pandey P, Singh S, Gupta R, Banerjee S (2015) Production of anti-cancer triterpene (betulinic acid) from callus cultures of different *Ocimum* species and its elicitation. Protoplasma 252:647–655

Papadopoulou K, Melton RE, Leggett M, Daniels MJ, Osbourn AE (1999) Compromised disease resistance in saponin-deficient plants. Proc Natl Acad Sci USA 96:12923–12928

Pu JX, Xiao WL, Lu Y, Li RT, Li HM, Zhang L, Huang SX, Li X, Zhao QS, Zheng QT, Sun HD (2005) Kadlongilactones A and B, two novel triterpene dilactones from *Kadsura longipedunculata* possessing a unique skeleton. Org Lett 7:5079–5082

Rastogi S, Kumar R, Chanotiya CS, Shanker K, Gupta MM, Nagegowda DA, Shasany AK (2013) 4-Coumarate: CoA ligase partitions metabolites for eugenol biosynthesis. Plant Cell Physiol 54:1238–1252

Rastogi S, Meena S, Bhattacharya A, Ghosh S, Shukla RK, Sangwan NS, Lal RK, Gupta MM, Lavania UC, Gupta V, Nagegowda DA, Shasany AK (2014) De novo sequencing and comparative analysis of holy and sweet basil transcriptomes. BMC Genom 15:588

Rastogi S, Kalra A, Gupta V, Khan F, Lal RK, Tripathi AK, Parameswaran S, Gopalakrishnan C, Ramaswamy G, Shasany AK (2015) Unravelling the genome of Holy basil: an "incomparable" "elixir of life" of traditional Indian medicine. BMC Genom 16:413

Reddy VA, Wang Q, Dhar N, Kumar N, Venkatesh PN, Rajan C, Panicker D, Sridhar V, Mao HZ, Sarojam R (2017) Spearmint R2R3-MYB transcription factor MsMYB negatively regulates monoterpene production and suppresses the expression of geranyl diphosphate synthase large subunit (MsGPPS.LSU). Plant Biotechnol J 15:1105–1119

Reed J, Stephenson MJ, Miettinen K, Brouwer B, Leveau A, Brett P, Goss RJM, Goossens A, O'Connell MA, Osbourn A (2017) A translational synthetic biology platform for rapid access to gram-scale quantities of novel drug-like molecules. Metab Eng 42:185–193

Robertson D (2004) VIGS vectors for gene silencing: many targets, many tools. Annu Rev Plant Biol 55:495–519

Ron M, Kajala K, Pauluzzi G, Wang D, Reynoso MA, Zumstein K, Garcha J, Winte S, Masson H, Inagaki S, Federici F, Sinha N, Deal RB, Bailey-Serres J, Brady SM (2014) Hairy root transformation using Agrobacterium rhizogenes as a tool for exploring cell type-specific gene expression and function using tomato as a model. Plant Physiol 166:455–469

Roy A, Saraf S (2006) Limonoids: overview of significant bioactive triterpenes distributed in plants kingdom. Biol Pharm Bull 29:191–201

Ruiz MT, Voinnet O, Baulcombe DC (1998) Initiation and maintenance of virus-induced gene silencing. Plant Cell 10:937–946

Sawai S, Saito K (2011) Triterpenoid biosynthesis and engineering in plants. Front Plant Sci 2:25

Shang Y, Ma Y, Zhou Y, Zhang H, Duan L, Chen H, Zeng J, Zhou Q, Wang S, Gu W, Liu M, Ren J, Gu X,

Zhang S, Wang Y, Yasukawa K, Bouwmeester HJ, Qi X, Zhang Z, Lucas WJ, Huang S (2014) Biosynthesis, regulation, and domestication of bitterness in cucumber. Science 346:1084–1088

Sheng H, Sun H (2011) Synthesis, biology and clinical significance of pentacyclic triterpenes: a multi-target approach to prevention and treatment of metabolic and vascular diseases. Nat Prod Rep 28:543–593

Siddiqui BS, Aslam H, Ali ST, Begum S, Khatoon N (2007) Two new triterpenoids and a steroidal glycoside from the aerial parts of *Ocimum basilicum*. Chem Pharm Bull (Tokyo) 55:516–519

Silva MG, Vieira IG, Mendes FN, Albuquerque IL, dos Santos RN, Silva FO, Morais SM (2008) Variation of ursolic acid content in eight *Ocimum* species from northeastern Brazil. Molecules 13:2482–2487

Sohrabi R, Huh JH, Badieyan S, Rakotondraibe LH, Kliebenstein DJ, Sobrado P, Tholl D (2015) *In planta* variation of volatile biosynthesis: an alternative biosynthetic route to the formation of the pathogen-induced volatile homoterpene DMNT via triterpene degradation in Arabidopsis roots. Plant Cell 27:874–890

Srivastava S, Conlan XA, Adholeya A, Cahill DM (2016) Elite hairy roots of *Ocimum basilicum* as a new source of rosmarinic acid and antioxidants. Plant Cell Tiss Org Cult 126:19–32

Tada H, Murakami Y, Omoto T, Shimomura K, Ishimaru K (1996) Rosmarinic acid and related phenolics in hairy root cultures of *Ocimum basilicum*. Phytochemistry 42:431–434

Tava A, Scotti C, Avato P (2011) Biosynthesis of saponins in the genus Medicago. Phytochem Rev 10:459–469

Thimmappa R, Geisler K, Louveau T, O'Maille P, Osbourn A (2014) Triterpene biosynthesis in plants. Annu Rev Plant Biol 65:225–257

Torre S, Tattini M, Brunetti C, Guidi L, Gori A, Marzano C, Landi M, Sebastiani F (2016) De novo assembly and comparative transcriptome analyses of red and green morphs of sweet basil grown in full sunlight. PLoS ONE 11:e0160370

Tsubaki S, Sugimura K, Teramoto Y, Yonemori K, Azuma J (2013) Cuticular membrane of Fuyu persimmon fruit is strengthened by triterpenoid nanofillers. PLoS ONE 8:e75275

Upadhyay AK, Chacko AR, Gandhimathi A, Ghosh P, Harini K, Joseph AP, Joshi AG, Karpe SD, Kaushik S, Kuravadi N, Lingu CS, Mahita J, Malarini R, Malhotra S, Malini M, Mathew OK, Mutt E, Naika M, Nitish S, Pasha SN, Raghavender US, Rajamani A, Shilpa S, Shingate PN, Singh HR, Sukhwal A, Sunitha MS, Sumathi M, Ramaswamy S, Gowda M, Sowdhamini R (2015) Genome sequencing of herb Tulsi (Ocimum tenuiflorum) unravels key genes behind its strong medicinal properties. BMC Plant Biol 15(1)

Wang Q, Reddy VA, Panicker D, Mao HZ, Kumar N, Rajan C, Venkatesh PN, Chua NH, Sarojam R (2016) Metabolic engineering of terpene biosynthesis in plants using a trichome-specific transcription factor MsYABBY5 from spearmint (*Mentha spicata*). Plant Biotechnol J 14:1619–1632

Wendt KU, Poralla K, Schulz GE (1997) Structure and function of a squalene cyclase. Science 277:1811–1815

Xu R, Fazio GC, Matsuda SP (2004) On the origins of triterpenoid skeletal diversity. Phytochem 65(3):261–291

Xue Z, Xu X, Zhou Y, Wang X, Zhang Y, Liu D, Zhao B, Duan L, Qi X (2018) Deficiency of a triterpene pathway results in humidity-sensitive genic male sterility in rice. Nat Communication 9(1):604

Zhan X, Yang L, Wang D, Zhu JK, Lang Z (2016) De novo assembly and analysis of the transcriptome of *Ocimum americanum* var. pilosum under cold stress. BMC Genom 17:209

Zhang DM, Xu HG, Wang L, Li YJ, Sun PH, Wu XM, Wang GJ, Chen WM, Ye WC (2015) Betulinic acid and its derivatives as potential antitumor agents. Med Res Rev 35:1127–1155f

Ocimum Genome Sequencing—A Futuristic Therapeutic Mine

10

Shubhra Rastogi and Ajit Kumar Shasany

Abstract

Next-generation sequencing (NGS) platforms from the past decade are in the continuous efforts of changing the impact of sequencing on our current knowledge about plant genes, genomes, and their regulation. Holy basil (*Ocimum tenuiflorum* L. or *sanctum* L.) genome sequencing has also paved the path for deeper exploration of the medicinal properties of this beneficial herb making it a true 'elixir of life.' The draft genome sequence of the holy basil has not only opened the avenues for the drug discovery but has also widened the prospects of the molecular breeding for development of new improved plant varieties.

10.1 Introduction

Herbs with medicinal and aromatic properties are becoming very popular worldwide. India including many Asian countries stands to be a gallery of many medicinal plant variety resources (Dhawale and Ghyare 2016) because of their ancient, prosperous, and the most distinct cultural backgrounds related to the use of such plants. Among such highly important herbs, *Ocimum* ranks first because of its wide range of health-promoting benefits and sanative properties in traditional as well as in the modern pharmacological system (Bhasin 2012) and hence is regarded as royal herb (Sharma 2010) or a queen of herbs (Kayastha et al. 2014). *Ocimum* is an omnipresent herb which belongs to the family lamiaceae, and this whole genus corresponds to the common name, 'basil.' The genus *Ocimum* is highly variable and possesses wide genetic diversity at intra- and inter-species levels (Upadhyay et al. 2015). It includes many aromatic herbs and shrubs having diverse essential oils of tremendous palliative importance which finds its applications in pharmaceutical, modern perfumery, and food processing industry.

Basically, nine species of *Ocimum* are grown worldwide, viz. *O. tenuiflorum* L., *O. gratissimum* L., *O. basilicum* L., *O. micranthum* L., *O. kilimandscharicum*, *O. americanum* L., *O. campechianum* L., *O. citriodorum* L., and *O. minimum* L., three of which *O. americanum* L., *O. citriodorum* L., and *O. minimum* L. All the species of *Ocimum* are economically and medicinally important. These plants have been extensively used in systems of traditional medicine across the world, primarily in Ayurveda and traditional Chinese medicine. Various parts of the plant have been consumed as a home remedy to cure many diseases since ages. According to Ayurveda, basil leaves are very good expectorant

S. Rastogi · A. K. Shasany (✉)
Plant Biotechnology Division, CSIR-Central Institute of Medicinal and Aromatic Plants, Lucknow 226015, India
e-mail: ak.shasany@cimap.res.in

A. K. Shasany and C. Kole (eds.), *The Ocimum Genome*, Compendium of Plant Genomes, https://doi.org/10.1007/978-3-319-97430-9_10

to relief cough and cold as well as are also used to treat digestive disorder. It is also beneficial in soothing itchy and irritated eyes (Brien 2009). In Chinese traditional medicine, basil leaves are resorted to relief the irritation caused due to insect bites (Brien 2009). These 5000-year-old methods of treatment are not just hearsay but are being supported by modern research. It has been demonstrated by the recent scientific research that constituents of basil essential oil acquire strong anticancer, antioxidant, antimicrobial and antiviral, properties (Bhattacharyya 2013). Hence, basil can be venerated as God's recipe for the elixir of life.

According to recent findings, it is suggested that some very high-value secondary metabolites such as geraniol, linalool, linalyl, camphor, citral, methyl eugenol, eugenol, methyl cinnamate, methyl chavicol, urosolic acid, safrol, thymol, taxol can be obtained from different plant parts which may prove to be a basis for generation of new drugs (Khare 2007). The mentioned metabolites are of great importance in medicine, aroma, and cosmetic trade. Despite its incomparable importance in the traditional system of medicine and huge repository of phytochemicals, our understanding behind the molecular aspect of this medicinal plant is still limiting. Recently, a comparative transcriptome study between *O. basilicum* and *O. sanctum* was published (Rastogi et al. 2014) wherein many transcripts of phenylpropanoids in *O. basilicum* and of terpenoid biosynthesis in *O. sanctum* were recorded. Again to decipher the complete metabolic routes and the aspects affecting the regulation and the channeling of these pathways, complete chloroplastic and nuclear genomes were sequenced for the first, after obtaining the sequencing data from three NGS platforms using four libraries (Rastogi et al. 2015). This report was the first step to unveil the secrets of its therapeutic potential with the scientific validation to the traditional claims of its utility in diverse medicinal usage and hence can be served as a milestone work for the identification of unidentified genes involving in the downstream biochemical pathway of medically significant metabolite. Figure 10.1 shows the pathways attributed to the proteins predicted from scaffolds of *O. tenuiflorum* genome as predicted by KAAS server in the study by Rastogi et al. (2015) considering *Oryza sativa* and *Arabidopsis thaliana* as reference organisms. Of the total annotated proteins, only 2.53% belonged to the category 'Biosynthesis of secondary metabolites' while 2.83% to the 'Metabolism of Terpenoids and polyketides.' In order to identify the key genes behind the strong medicinal properties of

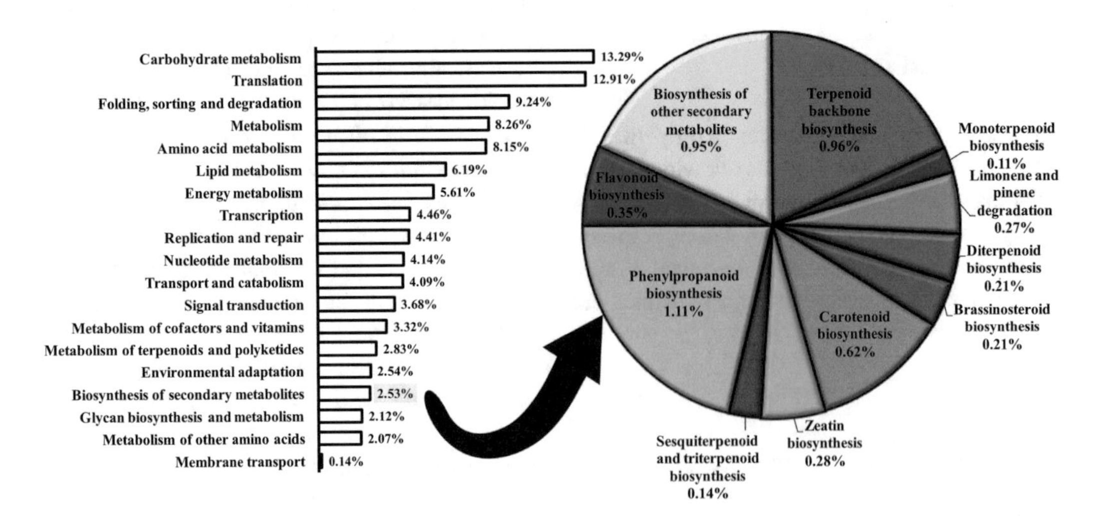

Fig. 10.1 Pathways related to the scaffolds predicted proteins by KAAS server considering *Oryza sativa* and *Arabidopsis thaliana* as reference organisms

this herb, another group (Upadhyay et al. 2015) also reported the draft genome of tulsi using Illumina HiSeq 1000 platform.

Figure 10.2 shows the top ten species that from which the predicted genes were annotated against all Viridiplantae clade genes in the UniProt database. Most annotations were made from the *Genlisea aurea* (order-Lamiales and family-Lentibulariaceae), i.e., 30,983 hits out of the total 53,480 annotated genes in UniProt. *G. aurea* is a carnivorous plant which is reported to be the plant with smallest known genome of size 63.6 Mb among the higher plants (Leushkin et al. 2013).

10.2 *Ocimum sanctum* (Syn. *tenuiflorum*): Whole Genome Sequencing

Modern sequencing techniques and bioinformatics approaches have presented better opportunities for rapid and efficient development of genomic resources for non-model plants such as *O. sanctum.* However, in contrast to the model plants and important crop plants, the projects on medicinal plant genomes are yet lagging behind. A speedy research is needed for the further development of natural medicines and high-yielding medicinal cultivars. In this context, recently CSIR-CIMAP, Lucknow, has succeeded in sequencing complete whole genome sequence of *O. sanctum.* Genome sequence of about 386 Mb from *O. sanctum* was assembled using whole-genome shortgun sequencing approach by generating short and long paired-end reads. Two libraries of long and short reads were of Illumina HiSeq 2000, one of SOLiD 550 XL and one of 454 GS FLX was combined to generate the draft genome assembly. Complete workflow of the sequencing process is provided in Fig. 10.3.

10.2.1 *O. sanctum* Whole Genome Sequencing: The Chloroplast and Mitochondrial Genome

This report by CSIR-CIMAP also revealed that chloroplastic genome of the *O. sanctum* is the smallest among lamiales with 142,524 bp in length challenging the investigation of Qian et al. (2013) who described the *Salvia miltiorrhiza* chloroplast genome to be the smallest (member of the *Ocimum* family Lamiaceae) with cp genome of length 151,328 bp (Table 10.1). In this analysis, *S. miltiorrhiza* was noticed to be phylogenetically nearest to *O. sanctum* with diploid chromosome no. $2n = 16$ (Rastogi et al. 2014). Both the plants predominantly produce

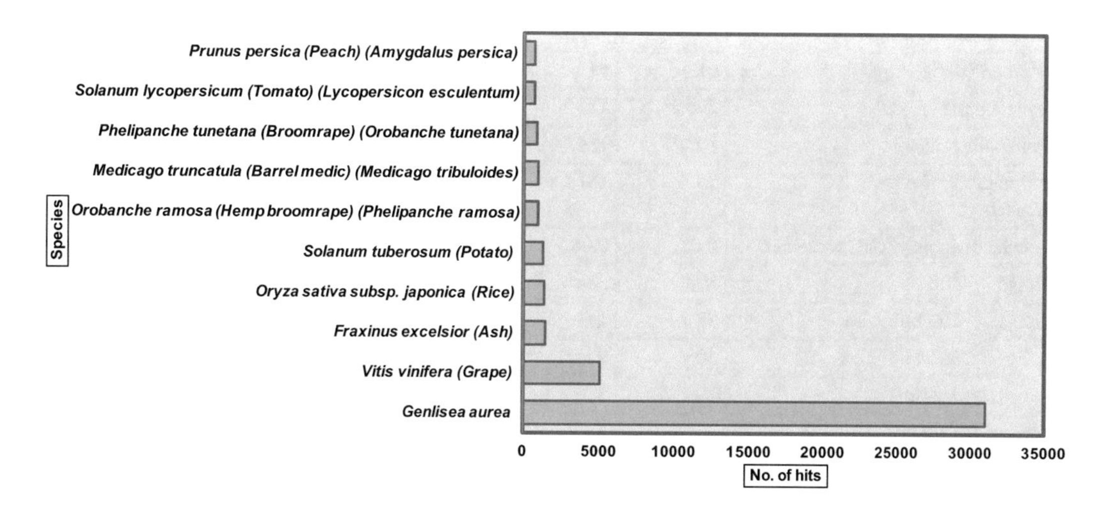

Fig. 10.2 Top 10 species of the predicted proteins from the viridiplantae clade in UniProt annotation table

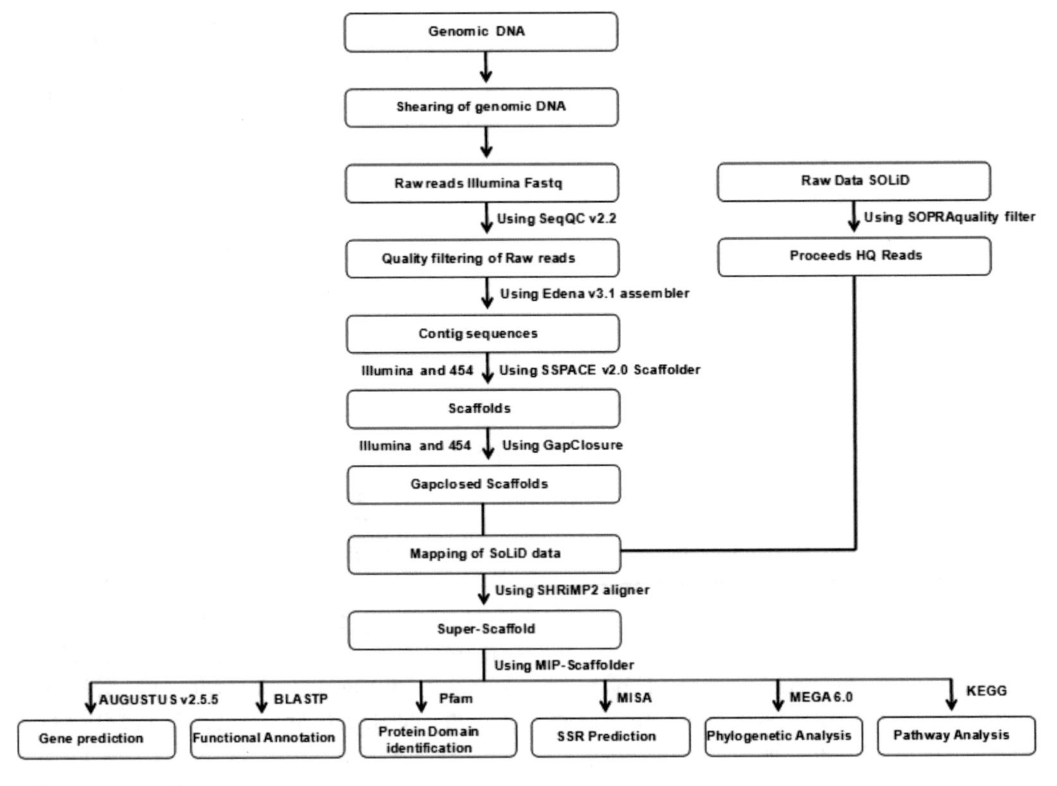

Fig. 10.3 Process workflow of *O. sanctum* whole genome sequencing and assembly

Table 10.1 QC statistics of chloroplast genome de novo assembly at each step

Particulars	Contigs	Scaffolds	Gap-closed	Gap-closed filtered	Draft genome
Contigs generated	140	48	48	2	1
Maximum contig length	25,631	78,214	78,166	78,166	1,42,524
Minimum contig length	61	61	61	64,356	1,42,524
Median contig length	296.5	133.5	70	71,261	1,42,524
Total contigs length	109,671	156,442	156,070	142,522	1,42,524
Total number of non-ATGC characters	55	1373	24	23	25
Percentage of non-ATGC characters	0.05	0.88	0.02	0.02	0.018
Contigs $\geq$ 100 bp	108	32	32	2	1
Contigs $\geq$ 200 bp	98	15	15	2	1
Contigs $\geq$ 500 bp	50	9	9	2	1
Contigs $\geq$ 1 Kbp	29	5	5	2	1
Contigs $\geq$ 10 Kbp	1	2	2	2	1
Contigs $\geq$ 1 Mbp	0	0	0	0	0
*N*50 value	1846	64,681	78,166	78,166	1,42,524

phenylpropanoids and their derivatives which are used in different traditional as well as in modern medicinal system. Like other asterid cp genomes, the overall GC content of *O. sanctum* was also found to be 36.2%. A total 158 genes from the cp genome of *O. sanctum* were reported that included 43 tRNA (transfer RNA) genes and 4 rRNA (ribosomal RNA) genes. Later in the same year, Upadhyay et al. (2015) published a reported discussing the draft genome of *O. tenuiflorum* L. (subtype *Krishna Tulsi*) using mate-pair and paired-end sequence library resulting 374 Mb draft genome. This report was published along with the comparative study of transcriptome of *O. tenuiflorum* subtype Rama and Krishna to find out the important differentially regulated gene and their role in the medicinally useful metabolite(s) synthesis. Both the reports published by Rastogi et al. (2015) and Upadhyay et al. (2015) are an important achievement in discovering the specialized plant metabolites of immense medicinal interest, glorifying the sacredness of the 'holy tulsi.' Chloroplasts have a vital function in supporting life on this planet. At present, there is an access to more than 800 chloroplast genomes which are sequenced from a range of land plants leading to expansion of our knowledge in chloroplast biology, gene transfer within the cells, preservation, diversity, and the basis of inheritance as a result of which chloroplast recombinant genes can be engineered for enhancement of agronomical plant traits or production of high-value farming or bio-scientific products. Since the chloroplast genome or the plastome exhibits a uniparental inheritance which is primarily maternal in angiosperms while paternal in gymnosperms, hence the sequence information coded in the chloroplast DNA loci is used in plant systematic for elucidating intraspecific relatedness via deep divergence (Martin et al. 2002; Kugita et al. 2003; Yamane et al. 2003; Ahmed et al. 2012, 2013). In some plant species, exogenous genes which encode important metabolites have also been transformed into the chloroplast genomes, like the genes for vaccines production effective against human diseases (Lössl and Waheed 2011; Waheed et al. 2011a, b). Particular loci in chloroplast DNA have also been used as barcodes which are helpful in the identification of plant species (Fazekas et al. 2008; Peter et al. 2009); however, the hypothesis of a universal barcode has natural precincts (Ahmed et al. 2013). Whole chloroplast genome sequence is considered to be a super-barcode (Li et al. 2015), but this notion requires reconsideration for outcrossing plant species that are, or have been occupying the same geographical range without loss of identity from interbreeding (sympatric species).

10.2.1.1 Some of the Biotechnological Applications of Chloroplast Genome Engineering

Production of Biomaterials and Enzymes

There are reports of the chloroplast genome engineering for the synthesis of important biomaterials, enzymes, and biofuels as well as biomass enhancement. Viitanen et al. (2004) for the first time reported the production of poly (*p*-hydroxybenzoic acid (pHBA) polymer at the highest level (25% dry weight) in normal healthy plants regardless of the alteration in the key metabolic intermediate via chloroplast genome engineering. Use of plant-derived enzyme cocktails from lignocellulosic biomass for the producing fermentable sugars was achieved for the first time by Verma et al. (2010). In contrast to the single biofuel enzymes which were formerly expressed in chloroplasts, nine discrete genes from heterologous systems (bacteria or fungi) were expressed in tobacco chloroplasts or *E. coli* by means of a novel method facilitating the introduction of fungal genes with several introns, eradicating the requirement of cDNA library preparation. Enzyme cocktails derived from the chloroplast propose several remarkable benefits like cost effectiveness, enhanced permanence of chloroplast-derived enzymes with no requirement of enzyme purification while the industry-based fermentation systems incorporate high expenditure and low manufacturing capability. Interestingly, β-glucosidase expression was able to release the hormones from conjugates, ensuing improved phytohormone levels and enhanced biomass (Jin et al. 2012), which was an unpredicted result of enzyme expression.

Improving Nutrition

Maize, soybean, and rapeseed (*Brassica napus*) seed oils are the main nutritional source of vitamin E. They acquire minimal α-tocopherol content but possess γ-tocopherol at fairly high levels. Only some oils from seed like sunflower seed oil (*Helianthus annuus*) are found to have high α-tocopherol content which is a key precursor leading to vitamin E biosynthesis (Schneider 2005). γ-Tocopherol leads to the synthesis of α-tocopherol catalyzed by γ-tocopherol methyl transferase (γ-TMT) and is thus the rate-limiting step (Shintani and DellaPenna 1998). When γ-tmt gene was overexpressed by engineering it into the chloroplast genome, it was observed that manifold layers were formed in the inner chloroplast envelope. In addition, approximately tenfold high production of α-tocopherol in the seed from γ-tocopherol was evident (Jin and Daniell 2014). Similarly, introduction of lycopene β-cyclase genes into the plastid genome of tomato improved the production of provitamin A (β-carotene) from lycopene, with apparent phenotypic modifications (Apel and Bock 2009).

Acquiring Stress Tolerance

For more than a decade, genetic engineering of chloroplast mainly focused on the over-expressing potential genes for increasing tolerance against biotic stress, which is imperative to plant defense and improving the yield. Insect pests are a real threat to the yield loss in many countries. Besides cotton bollworm resistance acquired by over-expressing Bt protein in chloroplasts (De Cosa et al. 2001), there are various other latest and prominent examples of enhanced biotic stress tolerance. Retrocyclin-101 and Protegrin-1 proteins were responsible for defending against tobacco mosaic virus (TMV) and Erwinia soft rot accountable to the yield deficit in a number of cultivated plants (Lee et al. 2011). β-glucosidase gene expression leads to the resistance against whitefly and aphid (Jin et al. 2011), which induces the liberation of insecticidal sugar esters by conjugating with hormones. Multiple defiances hostile to aphids, lepidopteran insects, whiteflies, and viral and bacterial pathogens were accomplished by expression of the Pinellia ternata agglutinin (PTA) gene in the genome of chloroplast (Jin et al. 2012). Edible crops like cabbage (*Brassica oleracea*) (Liu et al. 2007), soybean (Dufourmantel et al. 2004, 2005), and eggplant (*Solanum melongena*) (Singh et al. 2010) have been engineered and expressed stably by integrating over 40 transgenes in their chloroplast genomes leading to the improvement in the agronomic traits as well as the acquiring resistance against insects. As per the new strategy, scientists have now started exploring possibilities of downregulating the specific gene(s) of interest. Expressing dsRNAs (double-stranded RNAs) inside the genome of chloroplast to employ RNAi (RNA interference) for acquiring preferred agronomic traits, principally insect resistance in order to check yield loss is one such approach. This is achieved by expression of long or short dsRNAs activating the RNAi leading to the disruption of the target gene(s) in insects, thereby conferring effective protection from insects avoiding the use of harmful pesticide chemicals. This strategy was used by Jin et al. (2015) where they suppressed three vital proteins, namely Chi (lepidopteran chitin synthase), P450 (cytochrome P450 monooxygenase), and V-ATPase required for the survival of the insects by means of introducing dsRNAs in the chloroplast system of tobacco. Each of the double-stranded RNA was discretely expressed in chloroplasts, and leaves were then fed to the insects. The level of transcription in the target genes of Helicoverpa insects was found to be tremendously decreased to almost negligible in the midgut, ensuing considerable drop in the net larval weight as well as in the pupation rate (Jin et al. 2015). Transplastomic potato plants which produced β-actin and targeted long dsRNA were proved to be lethal to the larvae of Colorado potato beetle (*Leptinotarsa decemlineata*), demonstrating an additional crop protection method (Zhang et al. 2015).

Phyto-Pharmaceuticals

At this point in time, protein-based drugs are tremendously costly; for instance, insulin (a well-known drug for treating prevalent diabetes disease worldwide) is unaffordable to more than 90 percent of the world population. The excessive price of protein-based drugs is because of their expensive fermentation systems' setup (approximately $450–700 million based on their capacity) (Grabowski et al. 2006; Spök et al. 2008), exorbitant protein purification cost from the host, the requirement for low temperature storage and transfer, and the brief shelf-life of the ultimate product. Protein-based drugs prepared by plant chloroplasts defeat most of these challenges because there is no requirement of the costly fermentation systems and can easily be synthesized in hydroponic greenhouses approved by FDA (federal drug administration) (Holtz et al. 2015). Lettuce leaves which express protein drugs can be lyophilized and could be put in storage for an indefinite period at surrounding temperature without the loss in its efficacy (Su et al. 2015). Protein drugs encapsulated in the plant cell wall remain protected from the enzymes, and acids secreted by the human stomach as the plant cell wall glycans do not get digested by the human secreted enzymes. On the other hand, the microbes in the human gut have progressed in the breaking down each of the glycosidic bond present in the plant cell wall, consequently liberating the protein drug in the lumen of gut, guiding its transport to the immune system or blood (El Kaoutari et al. 2013; Kwon and Daniell 2015). Mode of oral release of various human curative proteins expressed in chloroplasts is has proven to be an effective medication for treating numerous human ailments, like diabetes, pulmonary hypertension, cardiovascular disease, and Alzheimer's disease. Majority of the proteins for the preliminary testing were expressed in chloroplasts of tobacco followed by their expression in chloroplasts of lettuce for subsequent expansion toward the clinic. Exendin-4 protein regulated the insulin secretion in a glucose-dependent way when delivered orally. Also the drop in glucose levels due to stimulation of insulin production in diabetic animals was observed, and this response was parallel to that generated by an injectable drug (Kwon et al. 2013). Oral intake of the bioencapsulated angiotensin-converting enzyme 2 (ACE2) and angiotensin (Ang) (1–7) enzymes leads to the better cardiopulmonary functioning, reduction in the high ventricular systolic blood pressure and recovered the blood flow in the animals with artificially generated pulmonary hypertension (Shenoy et al. 2014). Plant cell-encapsulated proteins, ACE2 and Ang (1–7), when delivered orally were also found to be reducing endotoxin induced uveitis (EIU) and significantly reduced retinal vasculitis and cellular infiltration, and additionally prevented folding and damage in induced autoimmune uveoretinitis animals (Kwon and Daniell 2015). Nonetheless, when the protein drugs were orally released to the Alzheimer's brain across the blood–brain barrier, it leads to the removal of plaques (Kohli et al. 2014). Su et al. (2015) for the first time reported the commercial-scale manufacturing of human blood-clotting factor (up to 30,000 doses for a 20-kg pediatric patient) in a 1000 ft 2 hydroponic cGMP facility. Lyophilized cells of clotting factor made in lettuce were found to last for the span of up to 2 years when stored at normal surrounding temperature, entirely removing the requirement of refrigerated storage. This leads to the advent of industrial level expansion of oral medicine addressing the challenges of high-priced purification, refrigerated storage and shipping, as well as the small shelf-life of the conventional protein drugs. Oral release of wider dose range proved to be efficient in the deterrence of antibody formation in response to clotting factor IX (FIX) injection, therefore making smooth the progress of human clinical studies.

Vaccines to Combat Infectious Diseases

The present iterated use of vaccines via attenuated strains of viruses and bacteria put forth safety against crucial transmittable diseases; however, they also offer major challenges. For instance, the oral polio vaccine which is used worldwide resulted in acute polio in consequence to the recombination and mutation other viruses (Chan and Daniell 2015). Besides, all existing vaccines demand refrigerated storage and shipping, making supply the major challenge in developing countries. Chloroplasts are the answer to such challenges. One vaccine derived from the chloroplast has been proved to be success granting double immunity against malaria and cholera in animal research (Davoodi-Semiromi et al. 2010). Cholera causes a high mortality rate, though there is only one approved vaccine which is not only high-priced but also has very small shelf-life. As far as malaria is concerned, there is no vaccine till date. In order to address this problem, cholera toxin-B subunit (CTB) of *Vibrio cholerae* was combined with the malarial vaccine antigen apical membrane antigen-1 (AMA1) and merozoite surface protein-1 (MSP1) for expression in lettuce or tobacco chloroplasts. Due to the unavailability of the human malarial model, the developed chloroplast-expressed CTB was tested by immunizing the mice and was found to be highly efficient. In addition, it also provided the highest duration of defense as per the literature (Davoodi-Semiromi et al. 2010). These initial results exhibit that chloroplasts are perfect model for production of inexpensive booster vaccines in counter to numerous contagious diseases (Lakshmi et al. 2013) for which the world is prepared; however, the deficient oral priming strategies yet are the major restraint in this area.

An attempt was also made to get the complete mitochondrial genome map of *O. sanctum* from the sequence assembly by considering *S. miltiorrhiza* as the reference mitochondrial genome instead of *Origanum vulgare* which was the reference chloroplast genome (Fig. 10.4). A total of 48 scaffolds from 140 contigs from cp genome, and 41 scaffolds from 124 contigs from the mitochondrial genome got generated (Table 10.2). Mitochondria play a significant role in the development of plant, maintaining its health as well as in plant reproduction. They are said to be semi-autonomous due to their own genetic system that makes it function for energy production, maintaining cell homeostasis and metabolism. Similar to other organisms, the plant mitochondrial genome determines a sequence of crucial polypeptides that put together the oxidative phosphorylation chain complexes with nuclear-encoded subunits. However, plant mitochondrial DNAs (mtDNAs) have distinguished characteristics that differentiate them from their fungal and animal counterparts. Particularly, higher plants contain large mtDNAs which greatly vary in sizes and structural make up. Furthermore, mitochondria in many plant species comprise a range of plasmids which are capable of independent replication. In comparison to animal mtDNA sequences, the majority of plant species harbor slow-evolving mtDNA gene sequences with an extremely low rate of point mutations. It is so due to the fact that mitochondrial DNA recombination system is very active and hence permits the mutations copy correction. Certainly, several investigations have demonstrated that mitochondrial genome of plants endures massive and highly frequent homologous recombination (HR). These events increase the rate of recurrence of rearrangements in plant mtDNA. Many characteristics of plant mtDNA genetics make it much complicated than most the other organisms. The ability of integration and/or expansion of intervening non-coding genetic sequences in the genomes of organelles is notable, particularly in comparison to the severely condensed mtDNA of mammals. It may therefore be hypothesized that advantage of sequences is associated to the competency of plant mitochondria for bringing in the DNA (Koulintchenko et al. 2003). Particularly, both the plasmid acquisition and the competency to import are common to mitochondria of fungus (Weber-Lot fi et al. 2009). On the other hand, mammalian organelles too are proficient (Koulintchenko et al. 2006) with stable genome size. Besides, many years later to the breakthrough, the importance of mitochondrial plasmids and their interaction with the core genome of mitochondria are yet obscure. It is not sequence content only, but the physical structure of the plant mitochondrial DNA is an intricate blend that results either from or by holds up an array of maintenance and replication method. Though

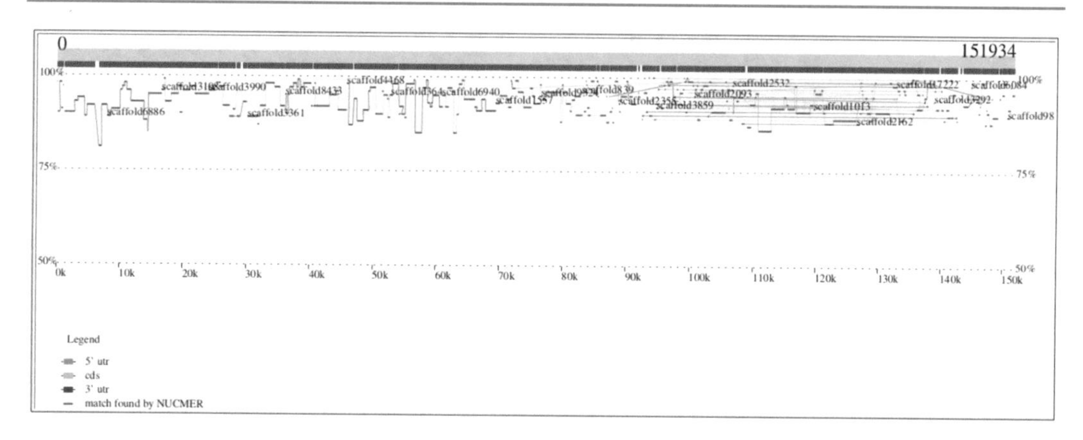

Fig. 10.4 Alignment of *O. sanctum* scaffolds with the chloroplast genome of *Origanum vulgare* using Nucmer

Table 10.2 QC statistics of mitochondrial genome de novo assembly at each step

Particulars	Contigs	Scaffolds	Gap-closed	Gap-closed filtered	Draft genome
Contigs generated	124	41	41	37	37
Maximum contig length	16,768	43,386	43,392	43,392	43,398
Minimum contig length	103	145	145	145	145
Median contig length	600	6261	505	524	20029
Total contigs length	174,923	442,572	442,370	446,661	4,45,881
Total number of non-ATGC characters	0	1616	51	4342	54
Percentage of non-ATGC characters	0	0.37	0.01	0.97	0.012
Contigs $\geq$ 100 bp	124	41	41	37	37
Contigs $\geq$ 200 bp	122	40	40	36	36
Contigs $\geq$ 500 bp	68	38	38	34	34
Contigs $\geq$ 1 Kbp	45	32	32	29	29
Contigs $\geq$ 10 Kbp	1	16	16	16	16
Contigs $\geq$ 1 Mbp	0	0	0	0	0
*N*50 value	3472	24,958	24,833	25,250	25,315

unusual in organelles of mammals, recombination plays a key role in determining plant mitochondrial genomes, reorganizing sequences, creating polymorphism, compelling evolution, and simultaneously conserving the genetic information. Eventually, challenge that stays behind is to know, that how could the high agility of the plant mitochondrial DNA be amalgamated with gene preservation, coordinating inter-compartment genome and apposite functional competence of the organelles. In this regard, it could be noticed that, despite reported mutants, mitochondrial DNA maps and restriction profiles have also been replicated persistently for a lot of species and generations, which is indicative of the fact that, even though vibrant and energetic, the genome of plant mitochondria remains constant or stable on an apparent timescale.

10.2.2 *O. sanctum* Whole Genome Sequencing: Re-routing the Traditional Health Practices to Scientific Drug Discovery

Traditional herbs have been used as medicines from the times immemorial, and *O. sanctum* (Tulsi) is one of the most indispensable ingredients of the traditional medicinal systems all round the globe. Despite this, the genetic makeup, the agricultural characteristics, as well as the curative properties of majority of conventional herbs are not properly understood. With speedy developments in the high-throughput sequencing methods and highly cut-down costs, a recent era of 'herbal genomics' has turned up. Research is now methodically classifying medicinal plants utilizing the next-generation sequencing approaches for annotation of their genomes for identifying and exploring the functions of their genes. The genomes of a few normally used medicinal herbs by now have been sequenced, like salvia (*Salvia miltiorrhiza*), tulsi (*O. sanctum*), lotus (*Lotus japonicus*), lingzhi (*Ganoderma lucidum*) (Sato et al. 2008; Chen et al. 2012; Rastogi et al. 2015; Xu et al. 2016). These medicinal plants could be presented as efficient model systems expanding the horizons for investigating secondary metabolites biosynthetic pathways in other medicinal plants. Genome data, in addition to the metabolomics, proteomic and transcriptomic information, could thus be helpful in predicting the biosynthetic pathways of secondary metabolites as well as their regulation. This would prove to be a reform in research leading to discovery fulfilling the goal to identify with the metabolic activities and genetics of medicinal plants.

Although, the modern healthcare system is developing day by day, but the emergence of many new diseases and the side effects of modern medicine, it is the call of an hour to integrate the traditional medical system for the ailment of lifestyle related health issues (Shakya 2016). *O. sanctum* is a model herbal plant used in the traditional Ayurvedic and Unani system of herbal medicine since ages (Pattanayak et al. 2010). It has incomparable health healing properties, and therefore is regarded as a 'boon' which provides health and longevity. Within few decades, bioactive compounds isolated from this plant may become the foundation for new pharmaceutical drugs. However, vague and scattered information of the ethnomedicines and unlinked databases to the other biomedical databases are the major obstacles for the development of phytomolecules isolated from *O. sanctum* into pharmaceutical drug with therapeutic efficacies. Therefore, the availability whole genome sequence of *O. sanctum* may help in developing few databases for sorting information concerned with single or multiple aspects of medicinal plants, like biologically active metabolites, ethnobotany, medicinal uses, information based on genome and or transcriptome, and molecular targeting active ingredients. Figure 10.5 shows the applications of *O. sanctum* genome sequence in pharmaceutical research based on traditional system.

10.2.3 *O. sanctum* Whole Genome Sequencing: A Tool for Molecular Identification and Elucidation of Novel Phytomolecules

O. sanctum produces a wide array of specialized metabolites of tremendous therapeutic potentials. Different plant part (leaf, stem, and root) contains a varying amount of many bioactive compounds which can be classified into several important classes on the basis of their synthesis in plants: terpenes (isoprenoids), polyketides, alkaloids, flavonoids, and phenylpropanoids. All of these classes contain many high-value chemicals whose genome sequence information is still unavailable. Identification of the biosynthetic gene involved in the production of such commercially important phytochemicals is often challenging. Prediction of the specific function to individual genes is a long and complex process because of limited genomic resources. For example, Orientin and Vicenin-2 are plant

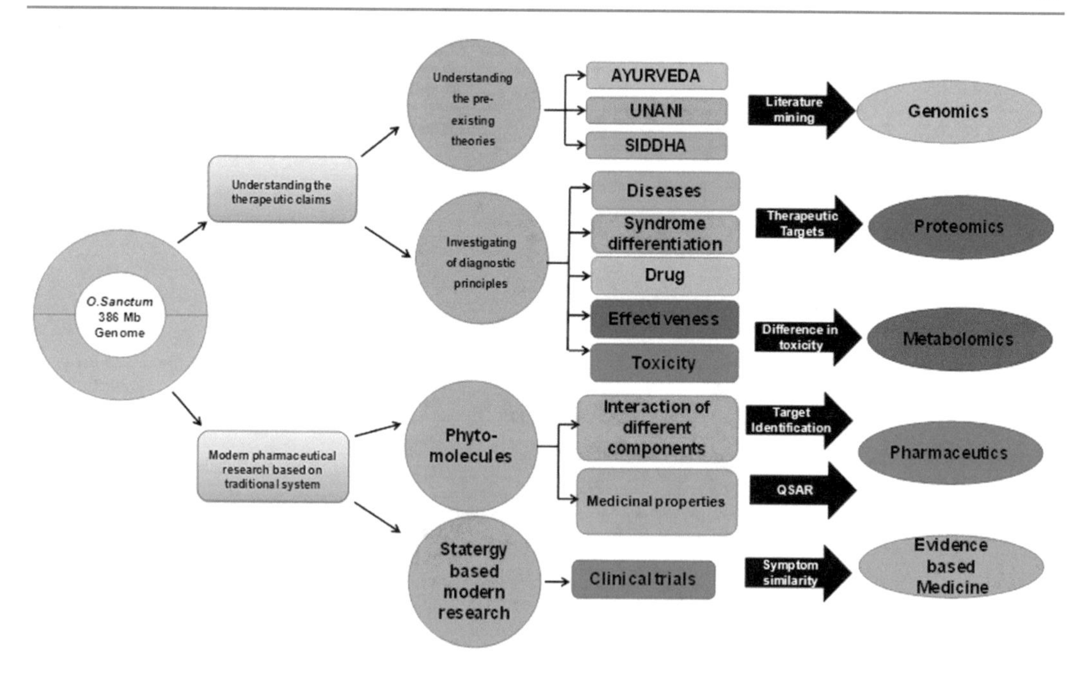

Fig. 10.5 Schematic diagram showing modern pharmaceutical research inspired by traditional medicine and the application of *O. sanctum* genome sequence in pharmaceutical research based on traditional system

flavonoid specifically isolated from the leaf extract of *O. sanctum* extensively studied for its antioxidant, anti-aging, antiviral, anti-inflammation, vasodilation, cardioprotective, radioprotective properties (Lam et al. 2016). Yet the underlying pathways and the mechanism of action of these flavonoid c-glycosides have remained indecisive. Similarly, putative genes and the network involved in the production of many other phytomolecules isolated from *Ocimum* such as ocimumosides, ocimarins, cirsimaritin, circineol are still unknown. Therefore, the availability of massive data in terms of the whole genome sequence of *O. sanctum* can be an efficient method for the investigation of the gene and gene families involved in the production these phytomolecules. Few examples of such compounds are illustrated in Table 10.3.

Also, data mining and the analysis of whole genome sequencing can be helpful in the identification of transcription factors and the response elements involved in their metabolite synthesis.

10.2.4 Functional Genomics Research in *Ocimum* Species

Though *Ocimum* species have a wide horizon in the food, flavor, and pharmaceutical industries, not much research has been carried out at the molecular genomic level. Table 10.4 reports the significant contribution of *Ocimum* genomics taking account of studies carried out at transcriptome, metabolome, and bioinformatics level leading to an assimilated knowledge of its natural products biosynthesis. But with the advent of draft genome of *Ocimum*, the sequences analysis may help in the identification of the small changes in amino acid at the substrate binding sites of metabolite biosynthesis pathway genes conferring unique therapeutic properties to this medicinal herb.

Beside the genomics, in the recent years, the research on in vitro methods of growing the *Ocimum* by the micropropagation and regeneration via *Agrobacterium*-mediated transformation

Table 10.3 List of some important plant molecules isolated from *O. sanctum*

Phytomolecule	Category	Structure	Therapeutic value	References
Ocimumoside A and Ocimumoside B	Glycoglycerolipids	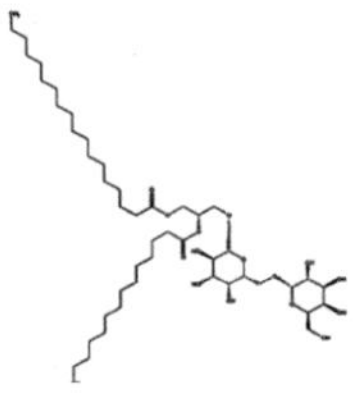	Used as an antistress agent	Ahmad et al. (2012)
Ocimarin	Coumarin	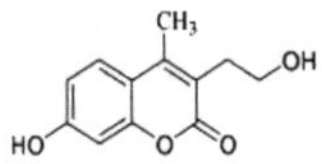	Used as an antistress agent	Gupta et al. (2007)
Cirsimaritin	Flavonoid	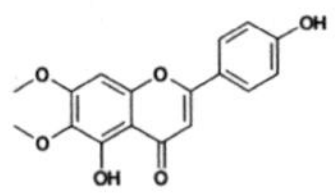	Possesses anti-inflammatory activity	Shin et al. (2017)
Cirsimarin	Flavonoid	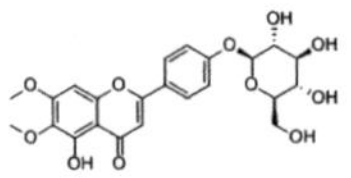	Used as a potent antilipogenic flavonoid to treat obesity	Zarrouki et al. (2010)
Orientin	Flavone-C-glycoside	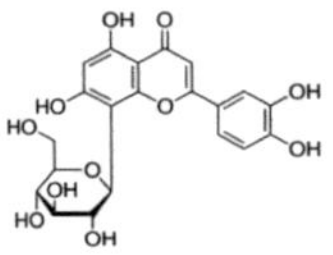	Used as an antioxidant, anti-aging, antiviral, antibacterial, anti-inflammation, vasodilatation and cardioprotective, radiation protective, neuroprotective, antidepressant-like, antiadipogenesis, and antinociceptive agent	Lam et al. (2016)
Vicenin-2	Flavone-C-glycoside	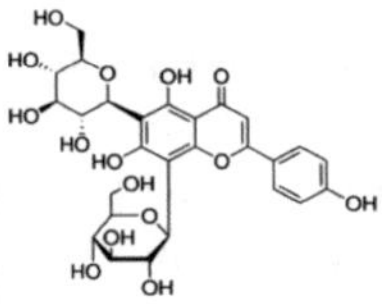	Used as a potential anti-inflammatory constituent	Marrassini et al. (2011)
Gardenin B	Flavone	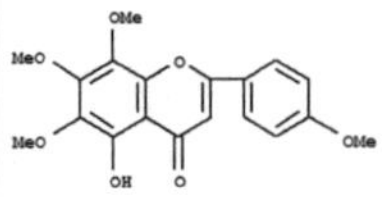	Possesses antiproliferative activity against the human leukemia cell	Cabrera et al. (2016)

Table 10.4 Functional genomics studies in *Ocimum* species

S.N.	Species	Gene(s)/biosynthetic pathway	Research	References
1.	*O. basilicum*	**Flavonoid *O*-methyltransferases** (flavonoid biosynthetic pathway)	Production of methoxylated flavonoids in yeast using ring A hydroxylases and flavonoid *O*-methyltransferases from sweet basil	Berim and Gang (2018)
2.	*O. kilimandscharicum*	**HMG-CoA reductase** (MVA pathway)	HMG-CoA reductase from camphor tulsi (*Ocimum kilimandscharicum*) regulated MVA-dependent biosynthesis of diverse terpenoids in homologous and heterologous plant systems	Bansal et al. (2018)
3.	*O. americanum*	**Pb1** (downy mildew resistance gene)	Transfer of downy mildew resistance from wild basil (*Ocimum americanum*) to sweet basil (*O. basilicum*)	Ben-naim et al. (2017)
4.	*O. basilicum*	**CYP716A (triterpenoid)** (terpene biosynthesis pathway)	Two CYP716A subfamily cytochrome P450 monooxygenases of sweet basil play similar but nonredundant roles in ursane- and oleanane-type pentacyclic triterpene biosynthesis	Misra et al. (2017)
5.	*O. basilicum*	**4-coumarate: CoA (4Cl), *p*-coumarate 3-hydroxylase (C3H0, caffeic acid *O*-methyltransferases (COMT), chavicol *O*-methyltransferase (CVOMT)** (phenylpropanoid biosynthesis pathway)	Water-deficit stress fluctuates expression profiles of 4Cl, C3H, COMT, CVOMT, and EOMT genes involved in the biosynthetic pathway of volatile phenylpropanoids alongside accumulation of methyl chavicol and methyl eugenol in different Iranian cultivars of basil	Khakdan et al. (2017)
6.	*O. basilicum*	MeJA responsive oxidosqualene cyclases **(*Ob*AS1 and *Ob*AS2)** (pentacyclic triterpene biosynthesis)	Methyl jasmonate-elicited transcriptional responses and pentacyclic triterpene biosynthesis in sweet basil	Misra et al. (2014)
7.	*O. basilicum*	Biosynthesis of flavonols and carotenoids	De novo assembly and comparative transcriptome analyses of red and green morphs of sweet basil grown in full sunlight	Torre et al. (2016)
8.	*O. basilicum*	**Thaumatin-like proteins (TLPs)** related to secondary metabolism	A thaumatin-like protein of *Ocimum basilicum* confers tolerance to fungal pathogen and abiotic stress in transgenic *Arabidopsis*	Misra et al. (2016)
9.	*O. americanum*	Targeted mainly on cold responsive genes	De novo assembly and analysis of the transcriptome of *Ocimum americanum* var. *pilosum* under cold stress	Zhan et al. (2016)

(continued)

Table 10.4 (continued)

S.N.	Species	Gene(s)/biosynthetic pathway	Research	References
10.	*O. gratissimum* *O. tenuiflorum* *O. kilimandscharicum Gürke* *O. americanum* *O. basilicum*	**Eugenol synthase** (phenylpropanoid biosynthetic pathway)	Comparative functional characterization of eugenol synthase from four different *Ocimum* species: implications on eugenol accumulation	Anand et al. (2016)
11.	*O. kilimandscharicum*	**Sesquiterpene synthase** (terpenoid biosynthesis pathway)	Functional characterization and transient expression manipulation of a new sesquiterpene synthase involved in β-caryophyllene accumulation in *Ocimum*	Jayaramaiah et al. (2016)
12.	*O. sanctum*	MEP/MVA pathway; phenylpropanoid biosynthesis pathway	Unraveling the genome of holy basil: an 'incomparable' 'elixir of life' of traditional Indian medicine	Rastogi et al. (2015)
13.	*Ocimum tenuiflorum*	Phenylpropanoid biosynthesis pathway/flavonoid biosynthesis pathway; terpenoid biosynthesis pathway	Genome sequencing of herb tulsi (*Ocimum tenuiflorum*) unravels key genes behind its strong medicinal properties	Upadhyay et al. (2015)
14.	*O. basilicum* *O. sanctum*	MEP/MVA pathway; phenylpropanoid biosynthesis pathway	De novo sequencing and comparative analysis of holy and sweet basil transcriptomes	Rastogi et al. (2014)
15.	*O. tenuiflorum*	**Eugenol *O*-methyltransferase (EOMT)** (phenylpropenoid biosynthesis pathway)	Characterization and functional analysis of eugenol *O*-methyltransferase gene reveal metabolite shifts, chemotype specific differential expression and developmental regulation in *Ocimum tenuiflorum* L.	Renu et al. (2014)
16.	*O. basilicum*	**Flavone 7-*O*-demethylase** (flavonoid biosynthesis pathway)	Identification of a unique 2-oxoglutarate-dependent flavone 7-*O*-demethylase completes the elucidation of the lipophilic flavone network in basil	Berim et al. (2014a, b)
17.	*O. basilicum*	**Flavones 8-hydroxylase (F8H)** **CYP82D33** (flavonoid biosynthesis pathway)	Unexpected roles for ancient proteins: flavone 8-hydroxylase in sweet basil trichomes is a Rieske-type, PAO-family oxygenase	Berim et al. (2014a, b)
18.	*O. sanctum*	**4-coumarate: CoA ligase (4Cl)** (phenylpropenoid biosynthesis pathway)	4-coumarate: CoA ligase partitions metabolites for eugenol biosynthesis	Rastogi et al. (2013)
19.	*O. basilicum*	**Flavone-6-hydroxylase (F6H) CYP82D monooxygenases** (flavonoid biosynthesis pathway)	The roles of a flavone-6-hydroxylase and 7-O-demethylation in the flavone biosynthetic network of sweet basil	Berim and Gang (2013)
20.	*O. basilicum*	MEP/terpenoid and shikimate/phenylpropanoid pathways	A systems biology investigation of the MEP/terpenoid and shikimate/phenylpropanoid pathways points to multiple levels of metabolic control in sweet basil glandular trichomes	Xie et al. (2008)

(continued)

Table 10.4 (continued)

S.N.	Species	Gene(s)/biosynthetic pathway	Research	References
21.	*O. basilicum*	**Cinnamate/p-coumarate carboxyl methyltransferases CCMT** (phenylpropenoid Biosynthesis pathway)	Evolution of cinnamate/ *p*-coumarate carboxyl methyltransferases and their role in the biosynthesis of methyl cinnamate	Kapteyn et al. (2007)
22.	*O. basilicum*	Methyl jasmonate	Effect of methyl jasmonate on secondary metabolites of sweet basil	Kim et al. (2006)
23.	*O. basilicum*	**Chavicol *O*-methyltransferase (CVOMT) Eugenol *O*-methyltransferases (EOMT)** (phenylpropenoid biosynthesis pathway)	Characterization of phenylpropene *O*-methyltransferases from sweet basil	Gang et al. (2002a, b)
24.	*O. basilicum*	*meta* hydroxylated phenylpropanoids **(phenylpropenoid biosynthesis pathway)**	Differential production of *meta* hydroxylated phenylpropanoids in sweet basil peltate glandular trichomes and leaves is controlled by the activities of specific acyltransferases and hydroxylases	Gang et al. (2002a, b)
25.	*O. basilicum*	**Caffeic acid *O*-methyltransferases (COMT)** (phenylpropenoid biosynthesis pathway)	Nucleotide sequences of two cDNAs encoding caffeic acid *O*-methyltransferases from sweet basil (*Ocimum basilicum*)	Wang et al. (1999)
26.	*O. basilicum*	Biosynthesis of phenylpropenes	An investigation of the storage and biosynthesis of phenylpropenes in sweet basil	Gang et al. (2001)

has also taken pace. Table 10.5 shows the record of in vitro studies in *Ocimum* species.

10.2.5 *O. sanctum* Whole Genome Sequencing: An Opportunity of Bringing Medicinal Plant into Cultivation

Today, about 80% of the world's population is dependent upon herbal medicine for their daily health care needs (Vines 2004). Hence, there is a threatening concern for a diminishing population of medicinal plants like *O. sanctum. O. sanctum* is one of the major medicinal plants used worldwide for its valuable herbal medicine. Infinite numbers of new complex, rare and bioactive compound are adding day by day to the list of important plant-based chemicals obtained from *O. sanctum*. There are still many such compounds needed to be explored which could be the foundation of many new pharmaceutical drugs synthesized artificially. There is an increasing demand of *O. sanctum* in pharmaceutical industry and is exploited regularly for its valuable compounds. This imposes a constant repression on the genetic diversity and its germplasm (Bhau 2012). With the advancement of DNA-based different molecular markers, conservation of biodiversity of medicinal plant could be supported. Molecular marker-based approaches can also be used for the medicinal crop improvement and molecular breeding strategies. Molecular markers are the unique DNA sequences used for determining genetic diversity among the individual or a population. These DNA sequences may be helpful in the

Table 10.5 In vitro studies in *Ocimum* species

S.N.	Species	Research	References
1.	*Ocimum tenuiflorum*	Elicitation of phenylpropanoids and expression analysis of PAL gene in suspension cell culture of *Ocimum tenuiflorum* L.	Vyas and Mukhopadhyay (2017)
2.	*Ocimum basilicum*	Efficient adventitious shoot organogenesis on root explants of *Ocimum basilicum* L.	Fraj et al. (2017)
3.	*Ocimum gratissimum*	Plant tissue culture: an alternative for production of useful secondary metabolites	Sravanthi et al. (2016)
4.	*O. gratissimum*	*Agrobacterium tumefaciens*-mediated genetic transformation of *Ocimum gratissimum*: a medicinally important crop	Khan et al. (2015)
5.	*Ocimum sanctum*	Protocol establishment for multiplication and regeneration of 'holy basil' (*Ocimum sanctum* Linn). An important medicinal plant with high religious value in India	Mishra (2015)
6.	*O. basilicum*	In vitro plant regeneration of Turkish sweet basil (*Ocimum basilicum* L.)	Ekmekci and Aasim (2014)
7.	*O. basilicum*	A calibrated protocol for direct regeneration of multiple shoots from in vitro apical bud of *Ocimum basilicum*—an important aromatic medicinal plant	Leelavathi et al. (2014)
8.	*O. sanctum*	An improved plant regeneration system of *Ocimum sanctum* L. —an important Indian holy basil plant	Sharma et al. (2014)
9.	*O. tenuiflorum*	Development of a rapid and high-frequency *Agrobacterium rhizogenes*-mediated transformation protocol for *Ocimum tenuiflorum*	Vyas and Mukhopadhyay (2014)
10.	*O. basilicum*	An efficient system for in vitro multiplication of *Ocimum basilicum* through node culture	Shahzad et al. (2012)
11.	*O. basilicum*	An efficient method for micropropagation of *Ocimum basilicum* L.	Saha et al. (2010)
12.	*O. sanctum*	Protocol establishment for multiplication and regeneration of *Ocimum sanctum* Linn. An important medicinal plant with high religious value in Bangladesh	Banu and Bari (2007)
13.	*O. basilicum*	Rapid micropropagation of *Ocimum basilicum* using shoot tip explants pre-cultured in thidiazuron supplemented liquid medium	Siddique and Anis (2007)
14.	*O. gratissimum*	In vitro multiplication of *Ocimum gratissimum* L. through direct regeneration	Gopi et al. (2006)
15.	*O. basilicum*	In vitro propagation of *Ocimum basilicum* L. (Lamiaceae)	Dode et al. (2003)
16.	*O. basilicum* and *O. citriodorum*	*Agrobacterium tumefaciens*-mediated transformation of *Ocimum basilicum* and *O. citriodorum*	Deschamps and Simon (2002)
17.	*O. basilicum*	Shoot regeneration of young leaf explants from basil (*Ocimum basilicum* L.)	Phippen et al. (2002)
18.	*O. basilicum*	In vitro rapid clonal propagation of *Ocimum basilicum*	Begum et al. (2002)
19.	*O. sanctum*	In vitro propagation of *Ocimum sanctum* L. through nodal explants. Bangladesh	Banu et al. (2001)
20.	*O. sanctum*	Micropropagation of 'holy basil' (*Ocimum sanctum* Linn.) from young inflorescences of mature plants	Singh and Sehgal (1999)

(continued)

Table 10.5 (continued)

S.N.	Species	Research	References
21.	*O. sanctum*	Callus induction and plant regeneration of *Ocimum sanctum*	Banu et al. (1999)
22.	*O. basilicum*	In vitro clonal propagation of an aromatic medicinal herb *Ocimum basilicum* L. (sweet basil) by axillary shoots proliferation	Sahoo et al. (1997)
23.	*O. americanum* *O. sanctum*	In vitro propagation of the medicinal herbs *Ocimum americanum* L. syn. *O. canum* Sims. (hoary basil) and *Ocimum sanctum* L. (holy basil)	Pattnaik and Chand (1996)

identification of homologous sequences in another related species (Canter et al. 2005). Whole genome sequencing of *O. sanctum* opens up an opportunity in the designing of new DNA markers from DNA sequences. Development of new DNA markers can be effectively utilized in integrating marker-based technology in conventional breeding.

According to Rastogi et al. (2015), simple sequence repeat (SSR) in the *O. sanctum* whole genome was predicted. A total of 4827 sequences which were greater than 500 bp length were studied for SSR and of total obtained sequences, 2612 possessed SSR repeats while the remaining (2364) sequences had more than one SSR (Fig. 10.6). The SSR markers obtained from *O. sanctum* can be utilized for the quantification of genetic diversity among different *Ocimum* genotype (Zietkiewicz et al. 1994; Dhawan et al. 2016). Prediction of other DNA-based markers from *O. sanctum* whole genome sequencing may bring advancement in breeding strategies as well as in resolving the problems faced during convectional breeding of *Ocimum* species. It could also be used for alteration of various traits among *Ocimum* species, for example; a cold tolerant gene from *O. kilimandscharicum* could be manipulated for the development of a cold tolerant *O. sanctum.* Similarly, many other genes from different biosynthetic pathways could be manipulated for the development of superior traits.

Further, gene editing tools such as CRISPR/Cas9 can also be utilized for simple manipulations and for the selection of an excellent trait of *O. sanctum*. It could find tremendous application in the identification of novel gene function and in pathway engineering of

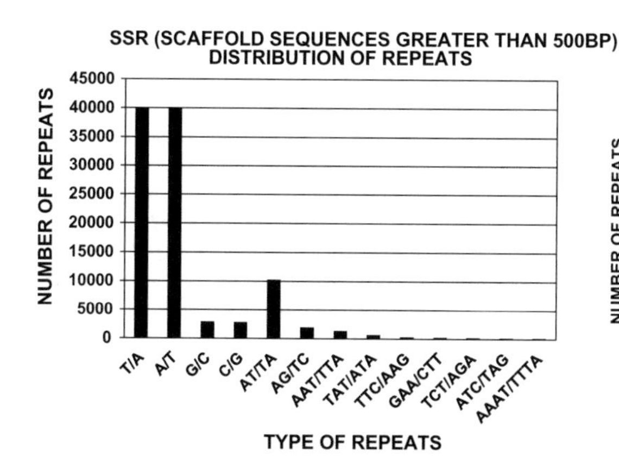

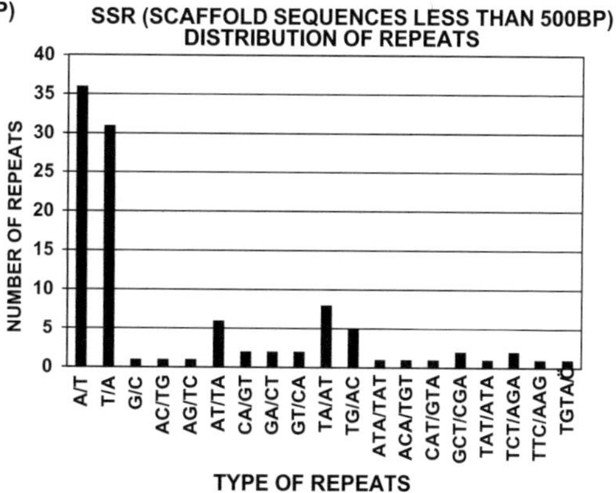

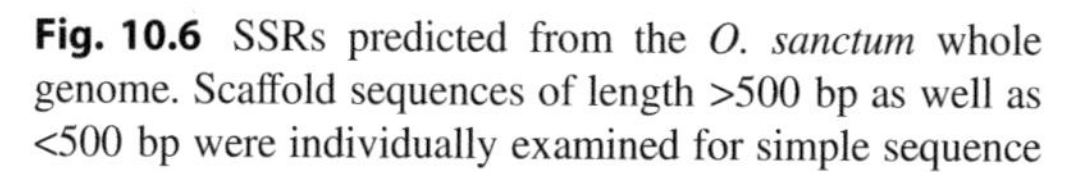

Fig. 10.6 SSRs predicted from the *O. sanctum* whole genome. Scaffold sequences of length >500 bp as well as <500 bp were individually examined for simple sequence repeats (SSRs) using MISA (http://pgrc.ipkgatersleben.de/misa/)

O. sanctum (Liu et al. 2017). However, it has to be done in a controlled manner and on a large scale. And it is only possible by bringing *O. sanctum* into cultivation.

10.3 Conclusion and Future Prospects

As compared to the animals, the genomes of plants are larger and intricate which code for the complex molecular systems those controlling the smooth functioning of the physiology. Due to the sessile nature of plants, their genomes had undergone evolution for generating an organism which adapts to the constant environmental changes. Additionally, in response to the theory of natural selection, the local wild-type crops have been sorted for preferred traits via plant breeding. Up to now, the selection breeding and the plant adaptability have produced plants that constitute several secondary metabolites. Nevertheless, global warming and the simultaneous diminution in the arable land, existing breeding approaches only would be unable in maintaining the endurance capacity or the exquisite metabolic composition of the plants. Hence, getting a comprehensive overview of the molecular networks operational in plants which regulate plant secondary metabolism is becoming essential in order to engineer those, as well as the detection of the novel genome-based markers for selecting such phenotypes would be critical to obtain the plants with specialized metabolic composition.

Availability of the whole genome sequence of *O. sanctum* will definitely accelerate the research of functional genes related to secondary metabolism. Such research can greatly contribute in the development of drug discovery based on natural product and acceptable exploitation of plant pharmaceutics sources. The genome sequence along with transcriptome sequence of *O. sanctum* can be greatly utilized for the genome re-sequencing which will be helpful in the detection of genetic molecular markers (GMMS), such as InDel, SSR, and SNP. Through the recent advances in NGS technologies, the genome sequencing of many targeted medicinal plant has now became possible because of reduced cost and lesser time utilized in the sequencing process. Sequencing and assembly of more and more medicinal plant will lead to the development of comparative genomics which would prove a widespread means to study the secondary metabolism of such therapeutic plants. Furthermore, studies on GMMS genetic linkage and the improved character, genotype, and phenotype of curative plant might play a major role in the molecular breeding of medicinal plants as well as in the development of the transgenics of such medicinal plants.

References

Ahmad A, Rasheed N, Gupta P, Singh S, Siripurapu KB, Ashraf GM, Kumar R, Chand K, Maurya R, Banu N, Al-Sheeha M (2012) Novel Ocimumoside A and B as anti-stress agents: modulation of brain monoamines and antioxidant systems in chronic unpredictable stress model in rats. Phytomedicine 19(7):639–647

Ahmed I, Biggs PJ, Matthews PJ, Collins LJ, Hendy MD (2012) Mutational dynamics of aroid chloroplast genomes. Genome Biol Evol 4:1316–1323

Ahmed I, Biggs PJ, Matthews PJ, Naeem M, Lockhart PJ (2013) Identification of chloroplast genome loci suitable for high-resolution phylogeographic studies of *Colocasia esculenta* (L.) Schott (Araceae) and closely related taxa. Mol Ecol Resour 13:929–937

Anand A, Jayaramaiah RH, Beedkar SD, Singh PA, Joshi RS, Mulani FA, Dholakia BB, Punekar SA, Gade WN, Thulasiram HV, Giri AP (2016) Comparative functional characterization of eugenol synthase from four different *Ocimum* species: implications on eugenol accumulation. Biochim Biophys Acta 1864(11):1539–1547

Apel W, Bock R (2009) Enhancement of carotenoid biosynthesis in transplastomic tomatoes by induced lycopene-to-provitamin A conversion. Plant Physiol 151:59–66

Bansal S, Narnoliya LK, Mishra B, Chandra M, Yadav RK, Sangwan NS (2018) HMG-CoA reductase from Camphor Tulsi (*Ocimum kilimandscharicum*) regulated MVA dependent biosynthesis of diverse terpenoids in homologous and heterologous plant systems. Sci Rep 8(1):3547

Banu LA, Bari MA (2007) Protocol establishment for multiplication and regeneration of *Ocimum sanctum* Linn. An important medicinal plant with high religious value in Bangladesh. Plant Sci 2(5):530–537

Banu LA, Bari MA, Islam R (1999) Callus induction and plant regeneration of *Ocimum sanctum*. In: Abstract. Proceedings of the 3rd international plant tissue culture conference, Dhaka, Bangladesh, 46

Banu LA, Bari MA, Haque E (2001) *In vitro* propagation of *Ocimum sanctum* L. through nodal explants. J Genet Biotechnol 2:143–146

Begum F, Amin MN, Azad MAK (2002) *In vitro* rapid clonal propagation of *Ocimum basilicum* L. Plant Tiss Cult 12:27–35

Ben-Naim Y, Falach L, Cohen Y (2017) Transfer of downy mildew resistance from wild basil (*Ocimum americanum*) to sweet basil (*O. basilicum*). Phytopathology 108(1):114–123

Berim A, Gang DR (2013) The roles of a flavone-6-hydroxylase and 7-O-demethylation in the flavone biosynthetic network of sweet basil. J Biol Chem 288 (3):1795–1805

Berim A, Gang DR (2018) Production of methoxylated flavonoids in yeast using ring A hydroxylases and flavonoid O-methyltransferases from sweet basil. Appl Microbiol Biotechnol. https://doi.org/10.1007/s00253-018-9043-0

Berim A, Kim MJ, Gang DR (2014a) Identification of a unique 2-oxoglutarate-dependent flavone 7-O-demethylase completes the elucidation of the lipophilic flavone network in basil. Plant Cell Physiol 56 (1):126–136

Berim A, Park JJ, Gang DR (2014b) Unexpected roles for ancient proteins: flavone 8-hydroxylase in sweet basil trichomes is a Rieske-type, PAO-family oxygenase. Plant J 80(3):385–395

Bhasin M (2012) *Ocimum*-Taxonomy, medicinal potentialities and economic value of essential oil. J Biosphere 1:48–50

Bhattacharyya P, Bishayee A (2013) *Ocimum sanctum* Linn. (Tulsi): an ethnomedicinal plant for the prevention and treatment of cancer. Anticancer Drugs 24 (7):659–666

Bhau BS (2012) Molecular markers in the improvement of the medicinal plants. Med Aromat Plants 1:e108. https://doi.org/10.4172/2167-0412.1000e108

Brien PO (2009) Basil in traditional Chinese medicine. How foods heal column, vol 7, pp 30–31. https://www.meridian-acupuncture-clinic.com/support-files/basil-in-tcm.pdf

Cabrera J, Saavedra E, del Rosario H, Perdomo J, Loro JF, Cifuente DA, Tonn CE, García C, Quintana J, Estévez F (2016) Gardenin B-induced cell death in human leukemia cells involves multiple caspases but is independent of the generation of reactive oxygen species. Chem-Biol Interact 256:220–227

Canter PH, Thomas H, Ernst E (2005) Bringing medicinal plants into cultivation: opportunities and challenges for biotechnology. Trends Biotechnol 23(4):180–185

Chan HT, Daniell H (2015) Plant-made oral vaccines against human infectious diseases—are we there yet? Plant Biotechnol J 13:1056–1070

Chen S, Xu J, Liu C, Zhu Y, Nelson DR et al (2012) Genome sequence of the model medicinal mushroom *Ganoderma lucidum*. Nat Commun 3:913

Davoodi-Semiromi A, Schreiber M, Nalapalli S, Verma D, Singh ND, Banks RK (2010) Chloroplast-derived vaccine antigens confer dual immunity against cholera and malaria by oral or injectable delivery. Plant Biotechnol J 8:223–242

De Cosa B, Moar W, Lee SB, Miller M, Daniell H (2001) Overexpression of the Bt cry2Aa2 operon in chloroplasts leads to formation of insecticidal crystals. Nat Biotechnol 19:71–74

Deschamps C, Simon J (2002) *Agrobacterium tumefaciens*-mediated transformation of *Ocimum basilicum* and *O. citriodorum*. Plant Cell Rep 21(4):359–364

Dhawale PG, Ghyare BP (2016) Antimicrobial activity and preliminary phytochemical studies on *Blepharis repens* (Vahl) roth. J Nat Sci Res 6(3):2225–3186

Dhawan SS, Shukla P, Gupta P, Lal RK (2016) A cold-tolerant evergreen interspecific hybrid of *Ocimum kilimandscharicum* and *Ocimum basilicum*: analyzing trichomes and molecular variations. Protoplasma 253(3):845–855

Dode LB, Bobrowski VL, Braga EJB, Seixas FK, Schuch MW (2003) *In vitro* propagation of *Ocimum basilicum* L. (Lamiaceae). Acta Sci Biol Sci 25 (2):435–437

Dufourmantel N, Pelissier B, Garcon F, Peltier G, Ferullo JM, Tissot G (2004) Generation of fertile transplastomic soybean. Plant Mol Biol 55:479–489

Dufourmantel N, Tissot G, Goutorbe F, Garcon F, Muhr C, Jansens S et al (2005) Generation and analysis of soybean plastid transformants expressing *Bacillus thuringiensis* Cry1Ab protoxin. Plant Mol Biol 58:659–668

Ekmekci H, Aasim M (2014) *In vitro* plant regeneration of Turkish sweet basil (*Ocimum basilicum* L.). J Anim Plant Sci 24:1758–1765

El Kaoutari A, Armougom F, Gordon JI, Raoult D, Henrissat B (2013) The abundance and variety of carbohydrate-active enzymes in the human gut microbiota. Nat Rev Microbiol 11:497–504

Fazekas AJ, Burgess KS, Kesanakurti PR, Graham SW, Newmaster SG et al (2008) Multiple multilocus DNA barcodes from the plastid genome discriminate plant species equally well. PLoS ONE 3:e2802

Fraj H, Hannachi C, Werbrouck SPO (2017) Efficient adventitious shoot organogenesis on root explants of *Ocimum basilicum* L. Acta Hort 1187:89–92. https://doi.org/10.17660/actahortic.2017.1187.10

Gang DR, Wang J, Dudareva N, Nam KH, Simon JE, Lewinsohn E, Pichersky E (2001) An investigation of the storage and biosynthesis of phenylpropenes in sweet basil. Plant Physiol 125(2):539–555

Gang DR, Beuerle T, Ullmann P, Werck-Reichhart D, Pichersky E (2002a) Differential production of *meta* hydroxylated phenylpropanoids in sweet basil peltate glandular trichomes and leaves is controlled by the activities of specific acyltransferases and hydroxylases. Plant Physiol 130(3):1536–1544

Gang DR, Lavid N, Zubieta C, Chen F, Beuerle T, Lewinsohn E, Noel JP, Pichersky E (2002b) Characterization of phenylpropene O-methyltransferases from sweet basil: facile change of substrate specificity and convergent evolution within a plant O-methyltransferase family. Plant Cell 14(2):505–519

Gopi C, Sekhar YN, Ponmurugan P (2006) *In vitro* multiplication of *Ocimum gratissimum* L. through direct regeneration. Afr J Biotechnol 5(9):723–726

Grabowski H, Cockburn I, Long G (2006) The market for follow-on biologics: how will it evolve? Health Aff 25:1291–1301

Gupta P, Yadav DK, Siripurapu KB, Palit G, Maurya R (2007) Constituents of *Ocimum sanctum* with antistress activity. J Nat Prod 70(9):1410–1416

Holtz BR, Berquist BR, Bennett LD, Kommineni VJ, Munigunti RK, White EL (2015) Commercial-scale biotherapeutics manufacturing facility for plant-made pharmaceuticals. Plant Biotechnol J 13:1180–1190

Jayaramaiah RH, Anand A, Beedkar SD, Dholakia BB, Punekar SA, Kalunke RM, Gade WN, Thulasiram HV, Giri AP (2016) Functional characterization and transient expression manipulation of a new sesquiterpene synthase involved in β-caryophyllene accumulation in *Ocimum*. Biochem Biophys Res Commun 473 (1):265–271

Jin S, Daniell H (2014) Expression of gamma-tocopherol methyltransferase in chloroplasts results in massive proliferation of the inner envelope membrane and decreases susceptibility to salt and metal-induced oxidative stresses by reducing reactive oxygen species. Plant Biotechnol J 12:1274–1285

Jin S, Kanagaraj A, Verma D, Lange T, Daniell H (2011) Release of hormones from conjugates: chloroplast expression of beta-glucosidase results in elevated phytohormone levels associated with significant increase in biomass and protection from aphids or whiteflies conferred by sucrose esters. Plant Physiol 155:222–235

Jin S, Zhang X, Daniell H (2012) *Pinellia ternata* agglutinin expression in chloroplasts confers broad spectrum resistance against aphid, whitefly, Lepidopteran insects, bacterial and viral pathogens. Plant Biotechnol J 10:313–327

Jin S, Singh ND, Li L, Zhang X, Daniell H (2015) Engineered chloroplast dsRNA silences cytochrome p450 monooxygenase, V-ATPase and chitin synthase genes in the insect gut and disrupts *Helicoverpa armigera* larval development and pupation. Plant Biotechnol J 13:435–446

Kapteyn J, Qualley AV, Xie Z, Fridman E, Dudareva N, Gang DR (2007) Evolution of cinnamate/p-coumarate carboxyl methyltransferases and their role in the biosynthesis of methylcinnamate. Plant Cell 19 (10):3212–3229

Kayastha BL (2014) Queen of herbs tulsi (*Ocimum sanctum*) removes impurities from water and plays disinfectant role. J Med Plants Stud 2(2):1–8

Khakdan F, Nasiri J, Ranjbar M, Alizadeh H (2017) Water deficit stress fluctuates expression profiles of 4Cl, C3H, COMT, CVOMT and EOMT genes involved in the biosynthetic pathway of volatile phenylpropanoids alongside accumulation of methylchavicol and methyleugenol in different Iranian cultivars of basil. J Plant Physiol 218:74–83

Khan S, Fahim N, Singh P, Rahman LU (2015) *Agrobacterium tumefaciens* mediated genetic transformation of *Ocimum gratissimum*: a medicinally important crop. Indust Crops Prod 71:138–146

Khare CP (2007) Indian medicinal plants: an illustrated dictionary. Springer, New York. https://doi.org/10.1007/978-0-387-70638-2

Kim HJ, Chen F, Wang X, Rajapakse NC (2006) Effect of methyl jasmonate on secondary metabolites of sweet basil (*Ocimum basilicum* L.). J Agric Food Chem 54:2327–2332

Kohli N, Westerveld DR, Ayache AC, Verma A, Shil P, Prasad T et al (2014) Oral delivery of bioencapsulated proteins across blood–brain and blood-retinal barriers. Mol Therap 22:535–546

Koulintchenko M, Konstantinov Y, Dietrich A (2003) Plant mitochondria actively import DNA via the permeability transition pore complex. EMBO J 22:1245e1254

Koulintchenko M, Temperley RJ, Mason PA, Dietrich A, Lightowlers RN (2006) Natural competence of mammalian mitochondria allows the molecular investigation of mitochondrial gene expression Hum Mol Genet 15:143e154

Kugita M, Akira K, Yuhei Y, Yuko T, Tohoru M et al (2003) The complete nucleotide sequence of the hornwort (*Anthoceros formosae*) chloroplast genome: insight into the earliest land plants. Nucleic Acids Res 31:716–721

Kwon KC, Daniell H (2015) Low-cost oral delivery of protein drugs bioencapsulated in plant cells. Plant Biotechnol J 13:1017–1022

Kwon KC, Nityanandam R, New JS, Daniell H (2013) Oral delivery of bioencapsulated exendin-4 expressed in chloroplasts lowers blood glucose level in mice and stimulates insulin secretion in beta-TC6 cells. Plant Biotechnol J 11:77–86

Lakshmi PS, Verma D, Yang X, Lloyd B, Daniell H (2013) Low cost tuberculosis vaccine antigens in capsules: expression in chloroplasts, bio-encapsulation, stability and functional evaluation in vitro. PLoS ONE 8:e54708

Lam KY, Ling APK, Koh RY, Wong YP, Say YH (2016) A review on medicinal properties of orientin. Adv Pharmacol Sci 1–9. http://dx.doi.org/10.1155/2016/4104595

Lee SB, Li B, Jin S, Daniell H (2011) Expression and characterization of antimicrobial peptides Retrocyclin-101 and Protegrin-1 in chloroplasts to control viral and bacterial infections. Plant Biotechnol J 9:100–115

Leelavathi D, Kuppan N, Yashoda (2014) A calibrated protocol for direct regeneration of multiple shoots from in vitro apical bud of *Ocimum basilicum*—an important aromatic medicinal plant. J Pharm Res 8 (6):733–735

Leushkin EV, Sutormin RA, Nabieva ER, Penin AA, Kondrashov AS, Logacheva MD (2013) The miniature genome of a carnivorous plant *Genlisea aurea* contains a low number of genes and short non-coding sequences. BMC Genom 14:476

Li X, Yang Y, Robrt JH, Maurizio R, Yitao W et al (2015) Plant DNA barcoding: from gene to genome. Biol Rev 90:157–166

Liu CW, Lin CC, Chen JJ, Tseng MJ (2007) Stable chloroplast transformation in cabbage (*Brassica oleracea* L. var. *capitata* L.) by particle bombardment. Plant Cell Rep 26:1733–1744

Liu X, Wu S, Xu J, Sui C, Wei J (2017) Application of CRISPR/Cas9 in plant biology. Acta Pharmaceut Sin B 7(3):292–302

Lössl AG, Waheed MT (2011) Chloroplast-derived vaccines against human diseases: achievements, challenges and scopes. Plant Biotechnol J 9:527–539

Marrassini C, Davicino R, Acevedo C, Anesini C, Gorzalczany S, Ferraro G (2011) Vicenin-2, a potential anti-inflammatory constituent of *Urtica circularis*. J Nat Prod 74(6):1503–1507

Martin W, Rujan T, Richly E, Hansen A, Cornelsen S (2002) Evolutionary analysis of arabidopsis, cyanobacterial, and chloroplast genomes reveals plastid phylogeny and thousands of cyanobacterial genes in the nucleus. Proc Natl Acad Sci USA 99:12246–12251

Mishra T (2015) Protocol establishment for multiplication and regeneration of 'Holy Basil' (*Ocimum sanctum* Linn). An important medicinal plant with high religious value in India. J Med Plants Stud 3(4):16–19

Misra RC, Maiti P, Chanotiya CS, Shanker K, Ghosh S (2014) Methyl jasmonate-elicited transcriptional responses and pentacyclic triterpene biosynthesis in sweet basil. Plant Physiol 164(2):1028–1044

Misra RC, Kamthan M, Kumar S, Ghosh S (2016) A thaumatin-like protein of *Ocimum basilicum* confers tolerance to fungal pathogen and abiotic stress in transgenic *Arabidopsis*. Sci Rep 6:25340

Misra RC, Sharma S, Garg S, Chanotiya CS, Ghosh S (2017) Two CYP716A subfamily cytochrome P450 monooxygenases of sweet basil play similar but non-reduntant roles in ursane- and oleanane-type pentacyclic triterpene biosynthesis. New Phytol 214 (2):706–720

Pattanayak P, Behera P, Das D, Panda SK (2010) *Ocimum sanctum* Linn. A reservoir plant for therapeutic applications: an overview. Pharmacog Rev 4(7):95

Pattnaik S, Chand PK (1996) In vitro propagation of the medicinal herbs *Ocimum americanum* L. syn. *O. canum* Sims. (hoary basil) and *Ocimum sanctum* L. (holy basil). Plant Cell Rep 15(11):846–850

Peter MH, Laura LF, John L, Mehrdad H, Sujeevan R (2009) CBOL plant working group A DNA barcode for land plants. Proc Natl Acad Sci USA 106:12794–12797

Qian J, Song J, Gao H, Zhu Y, Xu J, Pang X, Yao H, Sun C, Li XE, Li C, Liu J (2013) The complete chloroplast genome sequence of the medicinal plant *Salvia miltiorrhiza*. PLoS ONE 8(2):e57607

Rastogi S, Kumar R, Chanotiya CS, Shanker K, Gupta MM, Nagegowda DA, Shasany AK (2013) 4-coumarate: CoA ligase partitions metabolites for eugenol biosynthesis. Plant Cell Physiol 54:1238–1252

Rastogi S, Meena S, Bhattacharya A, Ghosh S, Shukla RK, Sangwan NS, Lal RK, Gupta MM, Lavania UC, Gupta V, Nagegowda DA (2014) *De novo* sequencing and comparative analysis of holy and sweet basil transcriptomes. BMC Genom 15(1):588

Rastogi S, Kalra A, Gupta V, Khan F, Lal RK, Tripathi AK, Parameswaran S, Gopalakrishnan C, Ramaswamy G, Shasany AK (2015) Unravelling the genome of Holy basil: an "incomparable" "elixir of life" of traditional Indian medicine. BMC Genom 16 (1):413

Renu IK, Haque I, Kumar M, Poddar R, Bandopadhyay R, Rai A, Mukhopadhyay K (2014) Characterization and functional analysis of eugenol O-methyltransferase gene reveal metabolite shifts, chemotype specific differential expression and developmental regulation in *Ocimum tenuiflorum* L. Mol Biol Rep 41(3):1857–1870

Saha S, Ghosh PD, Sengupta C (2010) An efficient method for micropropagation of *Ocimum basilicum* L. Indian J Plant Physiol 15(2):168–172

Sahoo Y, Pattnaik SK, Chand PK (1997) *In vitro* clonal propagation of an aromatic medicinal herb *Ocimum basilicum* L. (sweet basil) by axillary shoots proliferation. In Vitro Cell Dev Biol-Plant 33(4):293–296

Sato S, Nakamura Y, Kaneko T, Asamizu E, Kato T, Nakao M, Sasamoto S, Watanabe A, Ono A, Kawashima K, Fujishiro T (2008) Genome structure of the legume, *Lotus japonicus*. DNA Res 15(4):227–239

Schneider C (2005) Chemistry and biology of vitamin E. Mol Nutr Food Res 49:7–30

Shahzad MA, Faisal M, Ahmad N, Anis M, Alatar A, Hend AA (2012) An efficient system for *in vitro* multiplication of *Ocimum basilicum* through node culture. Afr J Biotechnol 11(22):6055–6059

Shakya AK (2016) Medicinal plants: future source of new drugs. Int J Herb Med 4(4):59–64

Sharma T (2010) Toxic effect of *Ocimum sanctum* plant extract against *Acrida exaltata* (Orthoptera: acrididae) adults. J Env Res Dev 4(4):1008–1012

Sharma NK, Choudhary RC, Kumar M (2014) An improved plant regeneration system of *Ocimum sanctum* L.—an important Indian holy basil plant. J Cell Tiss Res 14(1):41–43

Shenoy V, Kwon KC, Rathinasabapathy A, Lin SN, Jin GY, Song CJ et al (2014) Oral delivery of angiotensin-converting enzyme 2 and angiotensin-(1–7) bioencapsulated in plant cells attenuates pulmonary hypertension. Hypertension 64:1248–1259

Shin MS, Park JY, Lee J, Yoo HH, Hahm DH, Lee SC, Lee S, Hwang GS, Jung K, Kang KS (2017) Anti-inflammatory effects and corresponding mechanisms of cirsimaritin extracted from *Cirsium japonicum* var. maackii Maxim. Bioorg Med Chem Lett 27 (14):3076–3080

Shintani D, DellaPenna D (1998) Elevating the vitamin E content of plants through metabolic engineering. Science 282:2098–3100

Siddique I, Anis M (2007) Rapid micropropagation of *Ocimum basilicum* using shoot tip explants pre-cultured in thidiazuron supplemented liquid medium. Biol Planta 51(4):787–790

Singh NK, Sehgal CB (1999) Micropropagation of ‘Holy Basil’ (*Ocimum sanctum* Linn.) from young inflorescences of mature plants. Plant Growth Regul 29 (3):161–166

Singh AK, Verma SS, Bansal KC (2010) Plastid transformation in eggplant (*Solanum melongena* L.). Transgenic Res 19:113–119

Spök A, Karner S, Stein AJ, Rodríguez-Cerezo E (2008) Plant molecular farming. Opportunities and challenges. JRC Sci Tech Rep 2008. http://ftp.jrc.es/EURdoc/JRC43873.pdf. Accessed 17 May 2016

Sravanthi M, Mohan GK, Suryakala G, Rani MS, Shanker K (2016) Plant tissue culture: an alternative for production of useful secondary metabolites. J Pharmacog Phytochem 5(4):269

Su J, Zhu L, Sherman A, Wang X, Lin S, Kamesh A (2015) Low cost industrial production of coagulation factor IX bioencapsulated in lettuce cells for oral tolerance induction in hemophilia B. Biomaterials 70:84–93

Torre S, Tattini M, Brunetti C, Guidi L, Gori A, Marzano C, Landi M, Sebastiani F (2016) *De novo* assembly and comparative transcriptome analyses of red and green morphs of sweet basil grown in full sunlight. PLoS ONE 11(8):0160370

Upadhyay AK, Chacko AR, Gandhimathi A, Ghosh P, Harini K, Joseph AP, Joshi AG, Karpe SD, Kaushik S, Kuravadi N, Lingu CS (2015) Genome sequencing of herb Tulsi (*Ocimum tenuiflorum*) unravels key genes behind its strong medicinal properties. BMC Plant Biol 15(1):212

Verma D, Kanagaraj A, Jin SX, Singh ND, Kolattukudy PE, Daniell H (2010) Chloroplast-derived enzyme cocktails hydrolyse lignocellulosic biomass and release fermentable sugars. Plant Biotechnol J 8:332–350

Viitanen PV, Devine AL, Khan MS, Deuel DL, Van Dyk DE, Daniell H (2004) Metabolic engineering of the chloroplast genome using the *Escherichia coli* ubiC gene reveals that chorismate is a readily abundant plant precursor for p-hydroxybenzoic acid biosynthesis. Plant Physiol 136:4048–4060

Vines G (2004) Herbal harvests with a future: towards sustainable sources for medicinal plants. Plantlife International. www.plantlife.org.uk

Vyas P, Mukhopadhyay K (2014) Development of a rapid and high frequency *Agrobacterium rhizogenes* mediated transformation protocol for *Ocimum tenuiflorum*. Biologia 69(6):765–770

Vyas P, Mukhopadhyay K (2017) Elicitation of phenylpropanoids and expression analysis of PAL gene in suspension cell culture of *Ocimum tenuiflorum* L. Proc Natl Acad Sci, India. Sect B: Biol Sci 1–11

Waheed MT, Thones N, Muller M, Hassan SW, Mona NR (2011a) Transplastomic expression of a modified human papillomavirus L1 protein leading to the assembly of capsomeres in tobacco: a step towards cost-effective second-generation vaccines. Transgen Res 20:271–282

Waheed MT, Thones N, Muller M, Hassan SW, Gottschamel J (2011b) Plastid expression of a double-pentameric vaccine candidate containing human papillomavirus-16 L1 antigen fused with LTB as adjuvant: transplastomic plants show pleiotropic phenotypes. Plant Biotechnol J 9:651–660

Wang J, Dudareva N, Kish CM, Simon JE, Lewinsohn E, Pichersky E (1999) Nucleotide sequences of two cDNAs encoding caffeic acid O-methyltransferases from sweet basil (*Ocimum basilicum*). Plant Physiol 120:1205

Weber-Lot fi F, Ibrahim N, Boesch P, Cosset A, Konstantinov Y, Lightowlers RN, Dietrich A (2009) Developing a genetic approach to investigate the mechanism of mitochondrial competence for DNA import. Biochim Biophys Acta 1787:320e327

Xie Z, Kapteyn J, Gang DR (2008) A systems biology investigation of the MEP/terpenoid and shikimate/phenylpropanoid pathways points to multiple levels of metabolic control in sweet basil glandular trichomes. Plant J 54(3):349–361

Xu H, Song J, Luo H, Zhang Y, Li Q (2016) Analysis of the genome sequence of the medicinal plant *Salvia miltiorrhiza*. Mol Plant 9(6):949–952

Yamane K, Yasui Y, Ohnishi O (2003) Intraspecific cpDNA variations of diploid and tetraploid perennial buckwheat, *Fagopyrum cymosum* (Polygonaceae). Am J Bot 90:339–346

Zarrouki B, Pillon NJ, Kalbacher E, Soula HA, N'jomen GN, Grand L, Chambert S, Geloen A, Soulage CO (2010) Cirsimarin, a potent antilipogenic flavonoid, decreases fat deposition in mice intra-abdominal adipose tissue. Int J Obesity 34(11):1566–1575

Zhan X, Yang L, Wang D, Zhu JK, Lang Z (2016) *De novo* assembly and analysis of the transcriptome of *Ocimum americanum* var. pilosum under cold stress. BMC Genom 17(1):209

Zhang J, Khan SA, Hasse C, Ruf S, Heckel DG, Bock R (2015) Pest control. Full crop protection from an insect pest by expression of long double-stranded RNAs in plastids. Science 347:991–994

Zietkiewicz E, Rafalski A, Labuda D (1994) Genome fingerprinting by simple sequence repeat (SSR)-anchored polymerase chain reaction amplification. Genomics 20(2):176–183

Printed in the United States
By Bookmasters